Spotlight SCIENCE 8

Keith JOHNSON   Sue ADAMSON   Gareth WILLIAMS

With the active support of: Lawrie Ryan, Bob Wakefield, Jan Green, Phil Bunyan, Jerry Wellington, Roger Frost, Kevin Sheldrick, Adrian Wheaton, Penelope Barber, John Bailey, Ann Johnson, Graham Adamson, Diana Williams.

Stanley Thornes (Publishers) Ltd

First published in 1994 by:
Stanley Thornes (Publishers) Ltd
Ellenborough House
Wellington Street
CHELTENHAM GL50 1YD
England
Reprinted 1994, 1995

A catalogue record for this book is
available from the British Library.

ISBN 0 7487 1562 2

Acknowledgements

The authors and publishers are grateful to the following for permission to reproduce photographs:

Adams Picture Library: 18BL, 121C, 124BR;
AEA Technology: 82L;
Allsport: 24, Bruno Gardent 63L, Gary M Prior 64T, 67, Stephen Munday 64B, Roger Labrosse 76T;
Heather Angel: 6TR, BR, 53L, CL, CR, R, 147;
Aquarius Picture Library: 70;
Ardea London: I.R. Beames 6CR, C&J Knights 11T, Valerie Taylor 12T;
A-Z Botanical Collection Limited: 13, 154T;
Barnaby's Picture Library: 108BL, 128;
Barts Medical Picture Library: 71B;
Bifotos/Soames Summerhays: 3T;
Biophoto Associates: 4(c), 12B, 74B, 158T;
British Airways: 35L;
British Steel: 115T;
Bubbles Photo Library: 29(4), 69B;
J. Allan Cash Photolibrary: 4(e), 6TCR, 7, 8B, 11BR, 17TL, 18C, 75, 102T, 108TR, TL, 119C, 120T, B, 121T, 148T;
Martyn Chillmaid: 9T, 18T, 22T, B, 23, 26T, B, 27TL, TR, TC, 29TR, TL, (1), (2), (3), (5), (6), (7), B, 34, 35R, C, 36R, L, 37R, L, 38, 39R, L, 40T, 42TL, TR, 76B, 83B, 91, 92R, C, 93T, 98T, 100T, 107, 110, 112B, 113, 114TL, TR, TCR, 116C, B, 120C, 122B, CL, 123, 125, 126T, BR, BC, BL, 127, 130T, 134, 135TL, TR, BL, BR, 139, 141C, B, 142T, CR, B, 143TL, BL, 146B, 151T, 152B, 157;
Bruce Coleman Limited: 54B, 62, 143TR, 154C, 156T, C, B;
Collections: 77;
Gene Cox: 148B, 152C, 153;
Department of Transport: 119T;
Ecoscene: 122T;
Eye Ubiquitous: 57, 59TR;
Fisons: 86BL;
Peter Fraenkel: 103C;
Leslie Garland Picture Library: 17CL;
GeoScience Features Picture Library: 94(c), 103BR, 104, 114C, 115B, 158B;
B.G. Grewar, Bridgemaster: 18BR;
Robert Harding Picture Library: 83TL, 97;
Holt Studios International: 9B, 86BR, 114TC, 118T, 150B, 152T;
Horticulture Research International: 150T, C;
Image Bank: L. Ternblad 42B, Michael Skott 43, Andy Caulfield 94(b), Mahaux Photography 105, Al Hamdan 119B, Hi-test Photo 121B;
Image Select/Ann Ronan Picture Library: 5, 27B;
Impact Photos: 122CR, Alain Le Garsmeur 124TL, TR;
The Frank Lane Picture Agency Ltd: 4(b), 50, 96C;
Magnum Photos: Chris Steele Perkins 143BR;
National Medical Slide Bank: 65;
National Meteorological Library: R. K. Pilsbury 94(d);
Oxford Scientific Films Ltd.: 4(a), (d), 6TL, TCL, BL, 10, 11BL, 69T, 71L, 92L, 108BR, 146T, 154B, 155T, B;
Panos Pictures: Neil Cooper 90, J. Hartley 96R, Trygve Bølstad 98C, B, Glenn Edwards 102B, Sean Sprague 103BL, Trevor Page 151B;
The Photographers' Library: 41, 54TR, 124BL;
Pilot Publishers Services Ltd.: 40B;
Quadrant Picture Library: 86T;
Raleigh: 17TR;
The Science Museum: 129;
Science Photo Library: 55T, B, 72, 73, Martin Dohrn 17CR, ESA/PLI 33, 51, 58 BL, Stevie Grand 40C, Sheila Terry 42TC, John Sanford 52, 54TL, NASA 58TL, 58TR, 59TL, 59BL, 60, US Geological Survey 58BR, Larry Mulvehill 63R, 74T, Alex Bartel 68, Dr KFR Schiller 79, Gordon Garradd 80, Stammers/Thompson 82R, 131, Martin Bond 93B, 96L, Angela Murphy 94(a), John Mead 145;
The Telegraph Colour Library: 8T, 30, ESA Meteosat 59CR, 83TR, 89, 103T, 116T, 141T;
Tony Stone Images 15, 17B, 32, 88, 99, 108CR, 111, 114B, TCL, 144, cover;
Topham Picture Source: 142CL;
Viewfinder: 112T, 124BC;
Wadworth & Co. Ltd.: 118B.

Typeset by Tech-Set, Gateshead,
Tyne & Wear.
Colour separations by Create Publishing
Services Limited, Bath.
Printed and bound in Spain by Mateu Cromo.

Contents

Find the key

Scientists use **keys** to identify living things.
A key has a number of questions.
You start at the beginning and answer "yes" or "no" to each question.
It will soon take you to the animal or plant you want.

▶ Use this key to identify these birds:

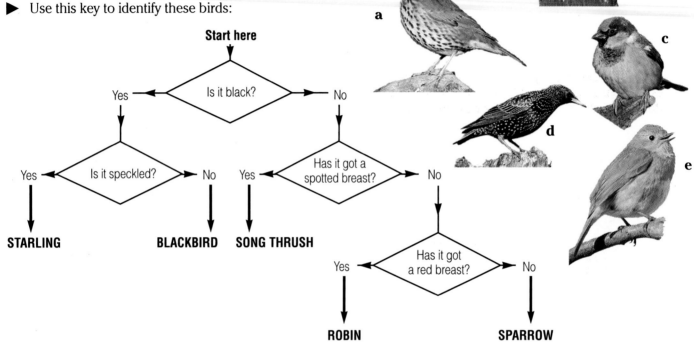

Start here

Is it black?
Yes ← → No

Is it speckled?
Yes ← → No

Has it got a spotted breast?
Yes ← → No

STARLING   **BLACKBIRD**   **SONG THRUSH**

Has it got a red breast?
Yes ← → No

ROBIN   **SPARROW**

a   b   c   d   e

Use the next key to identify 5 small animals found in grassland. It is set out differently from the first key, but works in the same way. Start at the beginning and answer the question at each stage.

1	Has legs	Go to 2
	Has no legs	Snail
2	Has 3 or 4 pairs of legs	Go to 3
	Has more than 3 or 4 pairs of legs	Centipede
3	Has 3 pairs of legs	Go to 4
	Has 4 pairs of legs	Spider
4	Has spots on body	Ladybird
	Has no spots on body	Ground beetle

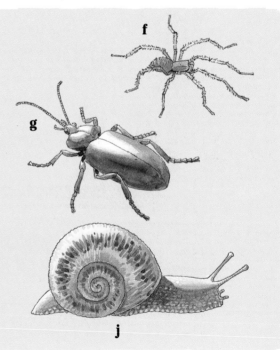

f   g   h   i   j

Leaf it out!

Now try making a key of your own.

1 Put the 6 leaves out in front of you.
2 Think of a question that divides them into 2 groups. Write the question down.
3 Now think up questions to divide each group into two. Write these down.
4 Carry on until you come to each leaf.
5 Write your key out neatly and then try it out on a friend.

▶ Now try making a key of these pond animals.
Do they have legs or not? What are their body shapes like?
Remember to split them up using one question at a time.

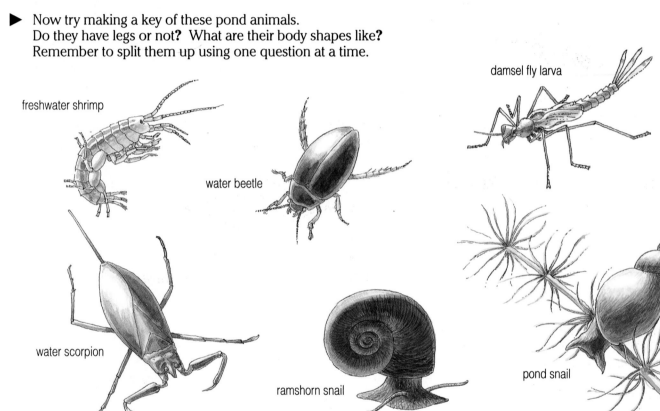

damsel fly larva

freshwater shrimp

water beetle

water scorpion

ramshorn snail

pond snail

Keys are good fun and get easier to use with practice.
Your teacher can give you more keys to try out.

Things to do

1 Cut out some photographs of animals or plants from old magazines. You could try dogs, cats, birds or flowers. Work out a key and make a poster to show how it works.

2 Write down a list of some things that belong in a group. The group could be pop singers or soccer teams or cars or maybe film stars. Make a key to use with your group.

3 Find 6 simple tools used in the kitchen. Work out a key to identify each one.

4 In the 18th century, Carl Linnaeus worked out a way of naming all living things. He gave them 2 names (called a **genus** and a **species**). His name for you is *Homo sapiens*. Find out what you can about Carl Linnaeus.

13b *Fit for survival*

Animals and plants have things about them that help them to survive in the place that they live.

▶ Look at the photographs.

Write down the things about each animal or plant that you think helps them to survive.

Oxford ragwort

polar bear

flounder

mole

eyed hawk moth

The animals and plants in the photographs are **adapted** to living in particular habitats.
They have special **adaptations** that help them to survive.

▶ List some of the adaptations that help you to survive.

Survival is the name of the game

Here are 2 animals that are adapted to live in harsh environments.

▶ List the adaptations that you think each animal shows. Then write down how each adaptation helps it to survive.

Mayfly larva

The mayfly larva lives in fast-flowing streams.

It clings tightly underneath rocks.

It has a flattened body with a very streamlined shape.

It feeds upon small plants growing on the rocks.

It has **gills** along the sides of its body and eyes on the top of its head.

It always moves away from the light into cracks between the stones.

Limpet

The limpet lives on the seashore.

It uses a sucker to cling tightly to rocks when the tide is out.

It has a thick shell to protect it from very high and very low temperatures.

It feeds on young seaweeds and it breathes using a gill to take in oxygen from the water.

Lice are nice!

Look carefully at your woodlice with a hand-lens.
Be careful not to damage them in any way.
How do you think they are adapted for living in leaf litter?

What conditions do you think woodlice like?
Make a list of your ideas (hypotheses).
You can use a **choice chamber** to find out if your ideas are right.

Plan an investigation to find out what conditions woodlice like.
Remember to make it a fair test.

- How many woodlice will you use?
- How will you record your results?

Show your plan to your teacher, then try it out.

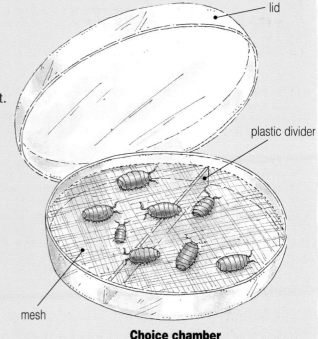

lid

plastic divider

mesh

Choice chamber

---

1 Why do you think that each of the following helps the animals to survive?
a) Deer and antelope are usually found together in herds.
b) Hoverflies have yellow stripes and look like wasps. But they are flies and have no sting.
c) Ragworms have good reflexes and can move back quickly into their burrows.

2 Suggest ways in which you think humans have been able to survive in the following environments:
a) hot desert   b) polar regions
c) highly populated cities.

3 Look at the numbers of eggs laid by these animals:

	Number of eggs
cod	3 million
frog	1000
snake	12
thrush	5

a) Why do you think the fish lays so many eggs?
b) Why does the snake lay far fewer eggs than the frog?
c) Why does a thrush lay so few eggs?

4 Look at these different bird beaks. Write down the name of each bird and say how you think its beak adapts it for survival.

Things to do

golden eagle

woodpecker   woodcock   wigeon

Competition

What does the word **competition** mean to you**?**

A race is a competition. Everyone tries very hard to win. But there can only be one winner.

In nature, living things compete for **resources** that are in short supply e.g. food and space.
Those that compete successfully will survive to breed.

▶ Write down some of the resources that animals compete for.

Write down some of the resources that plants compete for.

Seeing red

Robins compete for a **territory** (habitat) all the year round. They sing to let other robins know that the territory is occupied. During the breeding season they build a nest and raise their young inside the territory. At this time they are very fierce and drive other robins away.

a How many robin territories are shown on this map**?**

b What resources are the robins competing for**?**

c Why do you think that robins are so fierce to other robins but not to all birds that enter their territory**?**

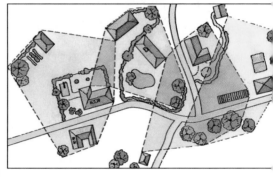

Weed this!

A weed is a plant that is growing where it is not wanted e.g. poppies in a wheat field.

It's easy to see why gardeners and farmers hate weeds and try hard to get rid of them.

Weeds compete with the other plants for light, water and space.

The dandelion is a successful weed.
Can you see why**?**

▶ Look at the adaptations of the dandelion in the diagram.

Copy and complete the table:

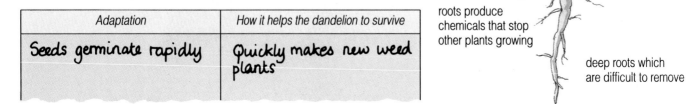

grows quickly and flowers twice a year

produces many seeds which are spread by the wind

resistant to many weedkillers

seeds germinate rapidly

leaves spread out over ground

grows quickly on bare soil

roots produce chemicals that stop other plants growing

deep roots which are difficult to remove

Adaptation	How it helps the dandelion to survive
Seeds germinate rapidly	Quickly makes new weed plants

Competition on the playing field

Dandelions, daisies and plantains compete with grass on your school field.

How could you find out which is the most successful weed?
You could count all of them, but this would take a very long time!
Instead you could take a **sample**. You could count the numbers of each weed in a small square called a **quadrat**.

1 Put your quadrat down on a typical area of the school field.
2 Count the numbers of dandelions, plantains and daisies inside your quadrat.
3 Take 4 more samples in different parts of the school field.
4 Record your results in a table like this:
- Add the totals of the class together for each weed.
- Draw a bar-chart of the class results.

d Why were you asked to take 5 samples?
e Which weed was the most successful on your school field?
f Try to think of a hypothesis that could explain this.
g What further investigation could you do to test this hypothesis?

Weed	Sample					
	1	2	3	4	5	Total
dandelions	3	3	4	0	2	12
plantains						
daisies						

Things to do

1 Copy and complete:
Living things . . . . for resources that are in . . . . supply, such as . . . . and . . . . Those plants and animals that . . . . successfully will . . . . to breed. Weeds compete with crops for . . . . and . . . .

2 Here are the planting instructions on a packet of broad bean seeds:
Sow the seeds about 5 cm deep and about 20 cm apart in open ground.
a) Why shouldn't the seeds be planted:
   i) any closer together? and
   ii) any further apart?
b) What resources might these plants compete for?

3 Can you think of any animals or plants that compete with humans?
Many of those that compete with us for food we call **pests**.
Write down any that you can think of and say what you think they compete with us for.

locust

A matter of life and death

Predators are animals that kill other animals for food.

The animals that they kill are called **prey**.

▶ Make a list of 5 animals that you think are predators.

a Predators are usually bigger and fewer in number than their prey. Why do you think this is?

Look at the tiger:
It seems to have some unbeatable weapons.

▶ Think about the things that make it a good predator. Make a list.

Sometimes we think that predators have an easy time killing defenceless prey.

In fact the tiger has to work hard for its meal. For every wild prey that it kills, the tiger fails 20–30 times.

b Predators often attack prey that is young, old, sick, weak or injured. Why do you think this is?

The predator strikes!

Choose one person to be the 'predator' in your group and blindfold him or her.

Arrange 9 discs at random on the squared paper.

Think of each disc as a prey animal.

Now the predator must search for the discs by tapping over the paper with one finger for one minute.

Each disc that the predator touches is removed. It counts as a 'kill'.

After each 'kill', the predator must pause and count to 3 before continuing.

After one minute count the total number of 'kills'.

Repeat the experiment, increasing the number of discs each time. Try 16, 25, 50 and 100 discs.

Record your results in a table.

Draw a graph of **number of kills** against **total number of prey**.

What are your conclusions from this experiment?

Don't get caught!

▶ Look at the hare:

c How is it adapted to escape predators**?**

Here are some things that help prey to survive.

Explain how each one increases the chances of their survival.

d Run, swim or fly fast.
e Stay together in large numbers.
f Taste horrible.
g Have warning colours.

Life's ups and downs

The graph shows the number of lynx (predators) and the number of hares (prey) over a number of years.

Look at the graph carefully.

If the hares have plenty of food they breed and they increase in number (see ①).

This makes more food for the lynx, so their numbers increase (see ②).

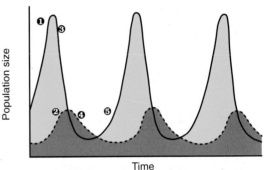

▶ Answer these questions about what happens next:

h Why do the numbers of hares fall at ③**?**
i Why do the numbers of lynx fall at ④**?**
j Why do the numbers of hares increase at ⑤**?**
k Why are the numbers of prey usually greater than the numbers of predators**?**

1 What do you think makes a good predator?
Make a picture of a make-believe predator that would catch lots of prey.

2 Some people think that predators are 'bad', but humans are the greatest predators that the world has known.
Write about some ways in which humans are predators.

3 Write about ways in which each of the following are successful predators:
a) domestic cat   b) spider   c) eagle.

4 All the animals preyed on by the wolf have very good ways of detection, defence and escape. The prey animals are usually safe from wolf attack.
So how is the wolf able to catch its prey?

Things to do

Population growth

A **population** is a group of the same animals or plants living in the same habitat e.g. greenfly on a rose bush or daisies in a lawn or a shoal of herring in the sea.

a Write down some more animal and plant populations and the habitats that they live in.

b Why do you think that animals and plants live together in populations?
Think about what they need for survival and how they keep up their numbers.

How do populations grow?

Simon put some yeast into sugar solution. It soon started to grow.

Every half hour he looked at a drop of the yeast under the microscope. He counted the number of cells he could see and recorded it in a table.

Then he drew a graph like the growth curve shown here:

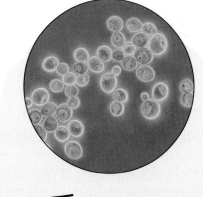

number of yeast cells

time (hours)

c What do you think happened to the number of yeast cells
i) in the first few hours?   ii) later in the experiment? and
iii) at the end of the experiment?

d Why do you think the population of yeast cells stopped growing?

When yeast grows one cell divides into 2. Then 2 become 4. Then 4 become 8, 16, 32, 64, and so on.

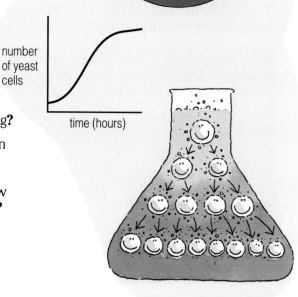

From Simon's experiment you can see that some populations grow very fast. So why isn't the Earth over-run with animals and plants?

The answer is that not all of them survive.
Some factors **limit** the population growth. For example,

- light
- overcrowding
- food and water
- disease
- climate
- predators
- oxygen
- shelter

▶ Copy and fill in the table for each factor above.

Factor	How the factor can limit population size
Light	lack of light slows growth of plants

What affects the growth of duckweed?

Duckweed is a small floating plant found in ponds.

If you put a single plant in a small beaker of water, it will grow and reproduce. Eventually it will cover the surface of the water.

Plan an investigation into the growth of duckweed.

Think about the factors that might affect the growth of the plant.

Choose one factor and investigate its effect.

- Plan your investigation to take about 3 weeks.
- How will you make it a fair test?
- How will you measure the amount of growth?
- How will you record your results?

Show your plan to your teacher before you try it out.

Human populations grow too!

The graph shows the increase in the world's population:

▶ Look at the graph and answer these questions:

e What do you think has caused the huge rise in the population over the last 300 years?

f Do you think that the number of people will continue to rise, level off, or fall?
Try to explain your answer.

g What sort of things might stop the human population from increasing?

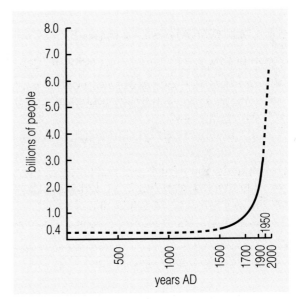

---

1 Copy and complete:
A . . . . is a group of animals or plants living in the same . . . . The growth of a . . . . can be limited by factors such as . . . . , ... and . . . . , so not all the young animals or plants in a population will . . . . to breed.

2 One of the slowest-breeding animals is the elephant. It has been worked out that, starting with one pair of elephants, their offspring would number 19 million after 700 years.
Explain why this could never actually happen.

3 Explain how you think each of the following would affect human population growth:
a) famine and disease
b) high birth rate
c) improved medical care.

Things to do

Questions

oxlip   corn poppy   welsh poppy

meadow saffron   wood garlic   marsh marigold

1 Look carefully at these wild flowers:
Make a key that you could use to identify each one.

2 A large herd of deer lived on an island.
The deer were sometimes killed by predators such as wolves.
To protect the deer population, some hunters shot all the wolves.
Then the deer population grew. Their numbers became so large
that they began to compete for grass. Many deer starved.
Soon the deer population was about the same as it was before the
wolves were shot.
a) What were the 4 populations involved?
b) Why do you think the hunters were wrong to shoot all the
   wolves?
c) What do you think will happen to the deer population in the
   future?

3 "Big, fierce animals are rare." Try to explain this statement.

4 Look at the way the leaves of these weeds are growing.
a) How do you think they survive trampling?
b) How do you think they affect the growth of the grass around
   them?
c) In what ways do gardeners cut down the competition from
   these weeds?

plantain   daisy   dandelion

5 Look at these samples of earthworm populations taken from 15 cm
under the soil:

Month	Jan.	Feb.	Mar.	Apr.	May	Jun.	Jul.	Aug.	Sep.	Oct.	Nov.	Dec.
number of worms	12	5	7	37	45	11	5	13	36	47	98	50
temperature (°C)	2	1	1	4	7	16	19	16	13	10	7	5
rainfall (mm)	40	30	25	50	80	20	5	25	40	50	80	70

a) Under what conditions do earthworms grow most successfully?
b) What do you think happens to the earthworms in hot weather?
c) Why do you think the earthworm numbers increase in the
   autumn?

greenfinch

6 Look at these pictures of birds' feet:
Write down the name of each bird and say
how their feet are adapted to help them
survive.

mallard   woodpecker   osprey

Using forces

14

Everything that you do needs a force – a push or a pull.

You have already investigated forces, using a spring-balance to measure them (in newtons).
You found out about weight and friction. You investigated floating and sinking. You measured the movement of toys.

In this topic you can use these ideas in new ways.

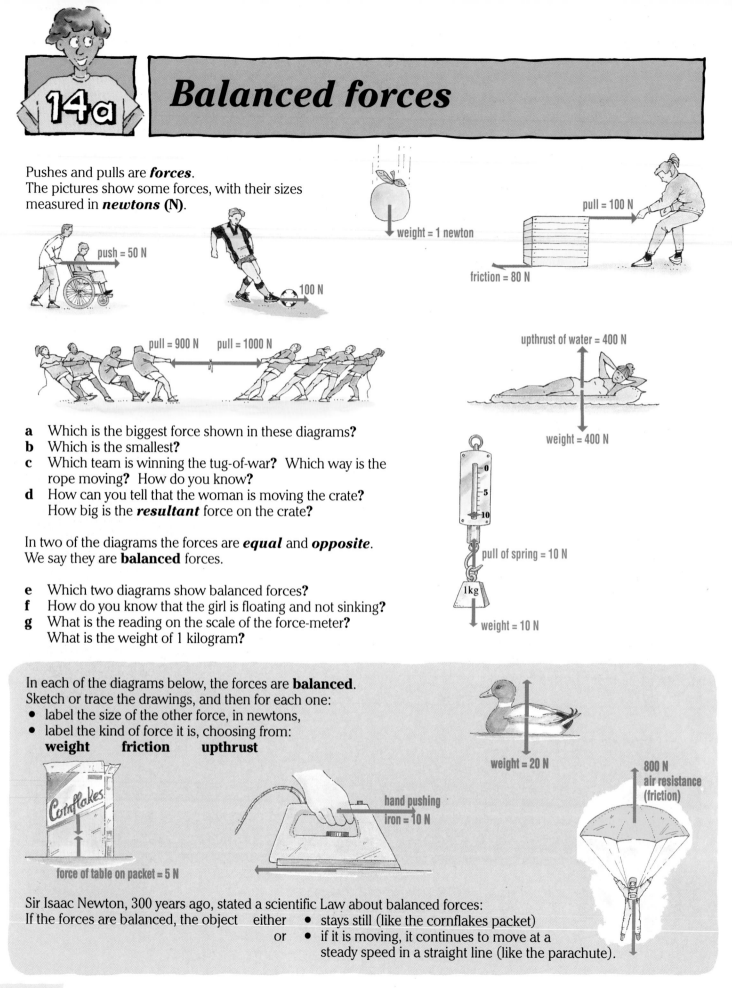

Balanced forces

Pushes and pulls are **forces**.
The pictures show some forces, with their sizes measured in **newtons (N)**.

push = 50 N

100 N

weight = 1 newton

pull = 100 N

friction = 80 N

pull = 900 N   pull = 1000 N

upthrust of water = 400 N

weight = 400 N

a   Which is the biggest force shown in these diagrams?
b   Which is the smallest?
c   Which team is winning the tug-of-war? Which way is the rope moving? How do you know?
d   How can you tell that the woman is moving the crate? How big is the **resultant** force on the crate?

In two of the diagrams the forces are **equal** and **opposite**.
We say they are **balanced** forces.

e   Which two diagrams show balanced forces?
f   How do you know that the girl is floating and not sinking?
g   What is the reading on the scale of the force-meter? What is the weight of 1 kilogram?

pull of spring = 10 N

1kg

weight = 10 N

In each of the diagrams below, the forces are **balanced**.
Sketch or trace the drawings, and then for each one:
• label the size of the other force, in newtons,
• label the kind of force it is, choosing from:
  weight   friction   upthrust

weight = 20 N

800 N
air resistance
(friction)

hand pushing
iron = 10 N

force of table on packet = 5 N

Sir Isaac Newton, 300 years ago, stated a scientific Law about balanced forces:
If the forces are balanced, the object   either   • stays still (like the cornflakes packet)
                                            or   • if it is moving, it continues to move at a
                                                      steady speed in a straight line (like the parachute).

Structures

Here are some photos of **structures**:

Sometimes you can see the structure.
For example, in a crane or a bridge, a fence or a tree, or a bicycle.

Sometimes the structure is hidden.
For example, the beams in the roof of your house, or the skeleton in your body.

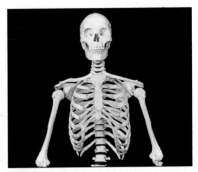

A structure can be designed by an engineer.
The structure must be strong enough to withstand the forces on it.
The forces in the structure must be **balanced** forces.

An engineering challenge!

Design and make a structure strong enough to support a 10 gram object as high as possible above the table.

You are given only:
- a 10 g object
- 20 straws
- 50 cm of sellotape. No more!

Your structure must be able to support the object for at least 30 seconds. Try it!

- Who can build the tallest successful tower?
- Draw a labelled sketch of your tower.
- Look at the highest towers: how many triangles of straws can you count? Triangular shapes help to make a structure firm and rigid.

The straws you used are hollow tubes. A tube is stronger than a solid bar of the same weight. Why is a bike made from tubes?
Tubes are found in animals (e.g. a bird's bones) and in plants (e.g. the stem of a dandelion).

---

1 Copy and complete:
a) Pushes and pulls are . . . .
b) When the forces on an object are equal and opposite, we say they are . . . .
c) Sir . . . . Newton's first Law is: if the forces on an object are . . . . , then
   - if the object is still, it stays . . . .
   - if the object is moving, it continues to . . . . at a steady . . . . in a . . . . line.
d) Structures are usually stronger if they are built of . . . . shapes.
e) A tube is . . . . than a solid bar of the same weight.

2 List all the structures that you can see in the classroom (or your home).

3 List all the structures that you can see on the way home.

4 Explain, with a diagram, why a bicycle frame is a strong structure.

5 Use the things you have learned to design a very thin tall tower for a TV transmitter. Draw a labelled diagram.

Things to do

Bridge the gap

▶ Bridges need to be safe and strong.
Look at this photo of a beam bridge:

a Write down 3 materials that a beam bridge could be made from.

b Name 3 materials that you would not use to build a bridge.

A bridge is a structure. In the structure some parts are being squashed.
This is called **compression**. The tiny particles are pushed closer together.

➡COMPRESSION⬅

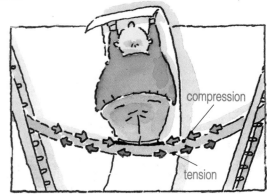

Other parts of the bridge are being stretched apart.
They are in **tension**.

⬅TENSION➡

c What do you think is happening to the tiny particles
where the beam is in tension?

d Bend your ruler gently. Which part is being stretched?
Draw a diagram of your bent ruler and label the parts that are in
tension and the parts in compression.

Here is a photo of a beam bridge made from girders in triangular shapes:

e Why are triangles used?

f Which part do you think is in tension?

Here are some more bridges.
Look at the way they are designed.

g Which parts do you think are: • in compression?
 • in tension?

An arch bridge

A suspension bridge

Investigating bridges

Investigation 1: Testing shapes

You are going to build some beam bridges, using only one sheet of A4 paper for each bridge.

You can use different designs for the beam. The picture will give you some help but use your own ideas as well.

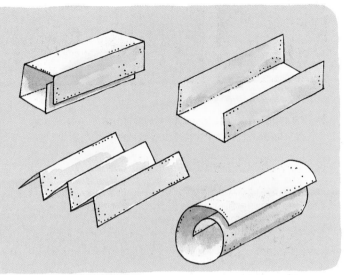

Plan an investigation to find out **which shape gives the strongest bridge**.

- How will you make it a fair test**?**

- Ask your teacher to check your plan and then do it.

- Record your results, and sketch the shapes you find are the best.

Investigation 2: Building a bridge

Your task is to build a bridge to cross a gap of 15 cm.
You can use straws, paper and sellotape – but they each have a cost.
Imagine that: 1 straw costs £1000
                    1 sheet of paper costs £1000
                    30 cm of sellotape costs £1000

Who can build the strongest bridge for £10 000?

- Design your bridge carefully. Make sure it has a roadway.

- Test it by adding weights until it collapses. Make sure you test your bridge in the same way as other people.
Be careful with the weights.

Write a report, including:
- a sketch of your design,
- how you tested it,
- where your design was weakest,
- how you could make it stronger.

---

1 When a simple beam bridge bends, the particles at the top are squashed together. This is called . . . .
Under the bridge, the bottom of the beam is in . . . ., and the . . . . are stretched apart.

2 Which kind of bridge (see the photographs) is best for bridging a wide river? Why?

3 Look at the photographs, and sketch simple diagrams of:
a) an arch bridge,
b) a suspension bridge.
On your sketches, colour in blue the parts you think are in compression, and colour in red the parts you think are in tension.

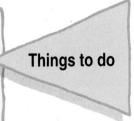

Things to do

Bending and stretching

14c

▶ Choose one of these 3 investigations, and do it.
If you have time, you can do a second one.

Bending beams

Tara and her family are painting the ceiling:

Tara sees that the plank sags when her father stands on it. It sags by a different amount when her baby sister stands on it.

- Write down what you think the sag of the plank depends on.
  Give as much detail as you can. (This is your hypothesis.)
  Try to include these words:

  weight   tension   compression   balanced forces

- Plan an investigation to see if your hypothesis is true.
  (You could use a ruler as the plank.)
  How will you record your results? ⚠

- Show your plan to your teacher, and then do it.

- When you have finished, write your report. Make sure you try to explain your results.

Stretching elastic

We often use elastic in our clothes – but sometimes it loses its 'stretchiness'.
Do you think its stretchiness changes when it is washed?

- Write down what you think happens to the stretchiness of elastic when it is washed. Do you think it depends on how it is washed? If so, in what way? (This is your hypothesis.)
  Try to include these words:

  force   tension   stretches   molecules

- Plan an investigation to see if your hypothesis is true.
  (Your teacher will give you some lengths of elastic.)
  How will you make it a fair test?
  How will you record your results?

  ⚠ eye protection

- Show your plan to your teacher, and then do it.

- When you have finished, write your report. Make sure you try to explain your results.

Stretching springs

Springs are useful in many ways:

- Make a list of all the uses of springs that you can think of.

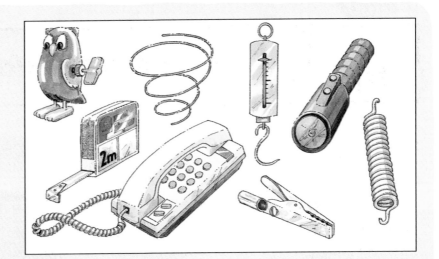

You can make your own spring by winding copper wire round a pencil:

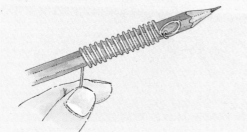

- Make a spring and then test it. Investigate how the length of your spring depends on the weight you hang on it.

- Plan your investigation carefully. How will you record your results?

- Check your plan with your teacher, and then do it. Start with small weights and carry on until your spring loses its shape.

- Plot a graph of the length (or the stretch) of your spring against the weight hanging on it. What do you notice?

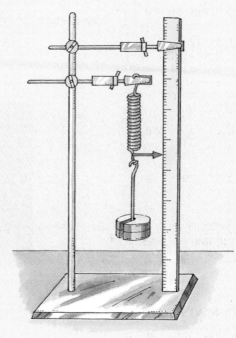

- If you have time, repeat your investigation with a spring made from iron wire or nichrome wire. What do you find?

- Write a report of what you did and what you found out.

---

1 Peter likes to go fishing. He tested 2 new fishing lines by hanging weights on them. Here are his results:

weight (N)	0	5	10	15	20
length of line A (cm)	50	51	52	53	54
length of line B (cm)	50	52	54	56	58

a) What happens to the lines as he hangs weights on them?
b) How long is line A when the load weight is 20 N?
c) What is the load when line B is 52 cm?
d) Which of the lines was the more stretchy?
e) Both lines are made of nylon. Which do you think will be the stronger one?

2 Amy has long blonde hair and Ben has short black hair. Design an investigation to compare their hair strength.

3 Design a pram with springs so that it can travel over rough ground without shaking the baby.

4 Robert Hooke investigated springs over 300 years ago and then wrote down Hooke's Law. Find out what is meant by Hooke's Law.

5 Use the data shown in Question 1 to plot 2 graphs (on the same axes).

Things to do

Moving at speed

Some things can move fast. Other things move slowly. They have a different **speed**.

Suppose the horse in the picture has a constant speed of 10 metres per second.
This means that it travels 10 metres in every second.

The speed can be found by:

Travelling at 10 m/s

$$\text{Average speed} = \frac{\textbf{distance travelled (in metres)}}{\textbf{time taken (in seconds)}}$$

Speed can also be measured in miles per hour (m.p.h.) or kilometres per hour (km/h).

Example
In a race, this girl runs 100 metres in 20 seconds.
What is her average speed?

Answer

$$\text{Average speed} = \frac{\text{distance travelled}}{\text{time taken}}$$

$$= \frac{100 \text{ metres}}{20 \text{ seconds}}$$

$$= 5 \text{ m/s}$$

(This is about 10 m.p.h.)

This is her **average** speed because she may speed up or slow down during the race.
If she speeds up, she is **accelerating**. If she slows down, she is **decelerating**.

▶ Copy out this table, and then complete it.

	Distance travelled	Time taken	Average speed
a	20 m	2 s	
b	100 m	5 s	
c	2 m		1 m/s
d		10 s	50 m/s
e	2000 km	2 h	

▶ Now match the speeds in the table with these objects. Which is which?
Add the names to the first column of your table.

Investigating speed

Investigation 1

Investigate how fast you can
- **walk**
- **run**

Plan your investigation first:

- What measurements will you take?
- How many measurements will you make?
- Check your plan with your teacher before starting.

Investigation 2

Chris says, **"I think people with long legs can always walk faster than people with short legs"**.

- Do you agree with this hypothesis?
- Plan a way to investigate this, and write a report explaining how you would do it.
- If you have time, check your plan with your teacher, and then do the investigation.

Investigation 3: A speed trap

Cars often drive fast past schools, and this is dangerous.

Plan an investigation **to find out how fast the cars are travelling on a road near your school**.

Do not do this investigation unless your teacher agrees it is safe.

- What distance will you use?
- How will you know when to start your clock accurately, and when to stop it?

If you do the investigation, find the speeds of 10 cars.

- What is the speed limit on this road?
- Were all the cars travelling within the speed limit?
- Discuss whether the speed limit should be higher or lower.
- Imagine you are a newspaper reporter. Use the results of your investigation to write a report for your newspaper.

30 m.p.h. is 13 m/s

---

1 Copy and complete:
a) The formula for speed is:
b) The units for speed are m/s (.... per ....) or km/h (.... per ....) or m.p.h. (.... per ....).
c) If a car speeds up it is ....
   If it slows down it is ....

2 Imagine you are in a car travelling through a town and then along a motorway. Give an example of where your car might have:
a) a high speed but a low acceleration,
b) a low speed but a high acceleration.

3 A boy jogs 10 metres in 5 seconds.
a) What is his speed?
b) How far would he travel in 100 seconds?

4 The table gives you some data about 4 runners:

Name	Distance (metres)	Time taken (seconds)
Ali	60	10
Ben	25	5
Chris	40	4
Dee	100	20

a) Who ran farthest?
b) Who ran for the shortest time?
c) Who ran fastest?
d) Which people ran at the same speed?

5 A dog runs at 10 m/s for a distance of 200 m. How long did it take?

Things to do

Questions

1 David is jumping out of an aeroplane:

a) When he first jumps out there is one force on him: his weight, which is 1000 N.
Draw a labelled diagram of this (with David drawn as a simple 'stick man').

b) Later his parachute opens, and he falls at a steady speed with 2 balanced forces on him: (i) his weight and (ii) air friction. Draw a labelled diagram of this.

c) After he lands and is standing still on the ground, there are 2 balanced forces on him: (i) his weight and (ii) the ground pushing up on his shoes. Draw a labelled diagram of this.

2 Oil tankers are very long ships. If the waves in the sea are high, they can lift a long ship so that it rests like a bridge between them:
Draw a sketch of this ship and then:

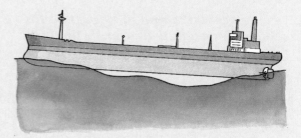

a) colour in blue the parts you think are in compression,

b) colour in red the parts you think are being stretched (in tension),

c) label your diagram.

3 Judy tested a spring by hanging weights on it. Here are her results:

a) Plot a line-graph of her results.

b) Write a sentence to say what conclusion you can draw from this graph.

weight (N)	1	2	3	4	5	6
extension (mm)	15	30	45	60	80	120

4 The data in the table is from the Highway Code. It shows the shortest stopping distances for a car at different speeds on a dry road. Look at the table carefully.

a) What do you think is meant by 'Thinking distance'?

b) How would it change if the driver was tired?

c) What do you think is meant by 'Braking distance'?

d) How would it change if the road was wet?

e), f), g) What are the missing numbers in the table?

h) What figures should be in the last row of the table?

Speed of car	Thinking distance (feet)	Braking distance (feet)	Overall stopping distance (feet)
20 mph	20	20	40
30 mph	(e)	45	75
40 mph	40	(f)	120
50 mph	50	125	(g)
60 mph			

Elements

Your life is full of elements.
You are made of them.
You eat them. You drink them.
You are surrounded by them.

In this topic you can find out more about elements.
What are they made of?
What can we make from them?

Chemical elements – the builders

▶ Write down your ideas about the following questions:

a How do you know that the air is all around you?

b What is passive smoking? Why do people worry about this?

c Something is cooking in the kitchen! Why can you smell it at the door?

In Book 7 you found out that everything is made from particles.
The particles are invisible. They are very small.
They are called **atoms**.
Atoms are the smallest parts of any substance.
They make up solids, liquids and gases.

▶ Look at this photograph of everyday substances:

Which do you think is the **simplest** substance?
Try to explain why.

Scientists have a name for the simplest substances.
They are called **elements**.
Elements are substances which cannot be broken down into anything simpler. **Elements** have **only one type of atom**.

You can show atoms like this:

oxygen atoms   All the atoms which make up oxygen gas are the same as each other.
Oxygen is an **element**.

nitrogen atoms   All the atoms which make up nitrogen gas are the same as each other.
Nitrogen is an **element**.

. . . **But** remember . . . oxygen atoms are different to nitrogen atoms.
Oxygen and nitrogen are **different** elements.

Your body is made up of many elements. Some of these are:

calcium	carbon	chlorine	hydrogen
magnesium	nitrogen	oxygen	phosphorus
potassium	sodium	sulphur	

Mostly these elements are not found on their own.
They are found in **compounds** in your body.
Compounds are substances which have 2 or more elements joined together. They have 2 or more different types of atom.

► Look at the names of some common compounds.
Which elements are they made from?

sodium chloride

carbon dioxide

hydrogen oxide (water)

Compounds look different to the elements they are made from.

sodium
(element)

+

chlorine
(element)

→

sodium chloride
(compound)

Compounds can also **behave** differently to the elements they are made from.
Think about some of the tests you could do to check this:

strength	hardness	stretchiness
density	melting point	solubility

Moon stones

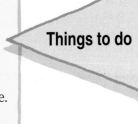

The moon stones are now on Earth!
You are the government's chief scientist.
Plan an investigation to look for similarities and differences between the solids.
Think about the tests you can do on elements and compounds.
Write detailed instructions for your assistants to do the tests.
Draw diagrams to help you to explain the tests.

If there is time, your teacher may let you try the tests.

Things to do

1 Copy and complete:
All substances are made of very small particles called . . . .
Substances which contain only one type of . . . . are called . . . .
. . . . cannot be broken down into anything simpler. When these combine together, they make . . . .

2 Look at labels on foods at home. Make a list of elements and compounds that the foods contain.

3 If you discovered a new element, what would you call it?

4 Say whether each of the following is an element or a compound:

chlorine	magnesium	iron
sulphur dioxide	sulphur	
carbon dioxide	iron chloride	
calcium	carbon	sodium

5 The first scientist to suggest the name element was Robert Boyle. The year was 1661. Find out some information about Boyle.

Compounds and mixtures

Elements have only one type of atom.
Compounds have 2 or more different atoms joined together.

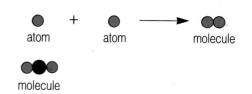

Which is the element? Which is the compound?

▶ Write down the names of 4 elements.
Write down the names of 4 compounds.

Atoms can join together. They make **molecules**.

2 oxygen atoms can join together. They make an oxygen **molecule**.
A carbon atom can join to 2 oxygen atoms. They make a **molecule** of carbon dioxide.
Notice that you can have molecules of an *element* or molecules of a *compound*.

atom + atom → molecule

molecule

Water is a **compound**. It is made from the **elements** hydrogen and oxygen.
You can use pictures to show this:

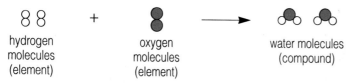

hydrogen molecules (element)   +   oxygen molecules (element)   →   water molecules (compound)

Why can't you *see* water molecules?

Spotting elements and compounds

Look at these diagrams of atoms and molecules.
Match the answers to the boxes.

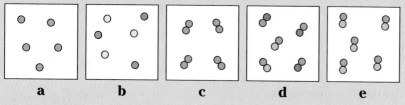

   a   b   c   d   e

i) Atoms of one element.
ii) Molecules of one element.
iii) Molecules of one compound.
iv) A mixture of 2 elements.
v) A mixture of 2 compounds.

We can divide all substances into 2 groups:

   pure substances   or   **mixtures**

A **pure substance** contains just one element or just one compound.
If you have a **mixture** you can separate it into pure substances.
Sometimes this is easy to do. Sometimes it is hard to do.

Could you sort out the sweets in this mixture?

Separating mixtures

Think about the following mixtures. Discuss them in your group. Say what experiments you would do to separate the parts of the mixture in each case. Ask your teacher for a Help Sheet if you get stuck!

Mixture 1 – A mixed bag!
Get the raisins from the mixed fruit.

Mixture 2 – An attractive problem!
Get the iron from the sulphur powder.

Mixture 3 – Mud in the garden!
Get dry soil from the water.

Mixture 4 – Swimming makes you thirsty!
Get pure water from salty water.

Mixture 5 – A broken sugar bowl!
Get the pieces of glass from the sugar.

Mixture 6 – Rain in the sand pit!
Get the dry sand *and water* from wet sand.

Mixture 7 – Colours unmixed!
Get the red dye from the purple mixture.

Your teacher may let you try some of these experiments.

Things to do

1 Use your own colours for atoms.
Draw 4 different boxes to show:
a) a mixture of 3 elements,
b) a pure compound,
c) a pure element,
d) a mixture of 2 compounds.

2 Draw a table like this:

Element	Mixture	Compound

Put these words in the correct columns.

magnesium air hydrogen water
salty water iron oxide chlorine
sulphur lemonade

3 Plan an investigation to see which of 3 liquids, A, B or C, evaporates fastest.

4 Match each word with its description.

Word	Description
element	made when atoms join together
molecule	one of 3 states of matter
evaporate	has only one type of atom
liquid	change from liquid to gas
condense	change from solid to liquid
melt	change from gas to liquid

5 Rock salt is salt from the ground. It contains particles of salt mixed with sand, dirt and other rocks.
Say how you would get a pure sample of dry salt from the rock salt.
Draw diagrams of any apparatus you would use.

Solving by separating

In the last lesson you found out some ways to separate mixtures. Can you remember some of them?

▶ The sentences **a** to **d** should tell you about separating mixtures. But the words are mixed up! Rearrange the words in each sentence so that it makes sense. Write out each correct sentence.

a insoluble Filtering separates liquid solid from.

b sea water from pure water Distillation gets.

c Chromatography coloured dyes a mixture in separates.

d separates magnet iron A from mixture a.

One of the most important mixtures to separate is crude oil. Crude oil is a fossil fuel. It is found underground. What does it look like?

Crude oil is *not* one pure substance. It is a **mixture** of many different chemicals. The chemicals are very useful to us.

▶ Make a list of some of the uses of crude oil. Life would be difficult without crude oil. Write about some of the things which would be most difficult *for you*.

Crude oil is separated by **fractional distillation**. The crude oil is heated to about 350 °C. At this temperature most of it **evaporates** (it turns to a gas). The hot mixture is put into a huge tower. It cools as it rises up the tower. The separate liquids in the mixture collect at different temperatures (boiling points). The boiling point is the temperature when the gas turns back to liquid.

e Is the tower hotter at the top or the bottom?

f What temperature does kerosine collect at?

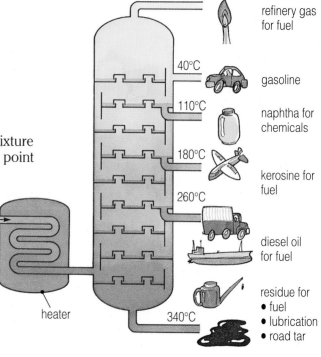

Fractional distillation can separate mixtures of other liquids. It works if the liquids have different boiling points.

Some liquids don't mix together. Another method is used to separate them. Think about oil and water.

▶ Design a piece of apparatus to separate oil and water. Label all the parts. Explain how it works.

You can see that crude oil contains lots of useful substances.

Being able to separate the substances in crude oil solves a big problem for the world. Can you be a problem-solver?

The people who have received these two memos need a bit of help. See if you can solve the problems using separating methods.

SMILEY NUTS KEEP A
SMILE ON YOUR FACE

MEMO TO: Lab Analysts
FROM: Managing Director

Smileynuts Ltd has been advertising the fact that its peanuts have a lower salt content than most other peanuts. As you know this is important for customers who have blood pressure problems.

I notice that 2 of our competitors have launched new low-salt brands — Crunchnuts and Nosalnuts. Please test Smileynuts and these two new nuts. I need to know which contains the least salt. Details of tests and results to me as soon as poss.

BANCO
Franchester Branch   PO Bo
PAY Cash
Three hundred p
only   E

From: The Chief

A forged cheque has been handed in to our station at Franchester. It has been written in ink from a fountain pen. Fingers Mike and Crafty Colin live in the area and our officers have found fountain pens in their possession. Please test the forged cheque ink and the ink from the pens belonging to Mike and Colin. Let me know the tests you do. Is either suspect guilty?

1   Which substance in crude oil is used to:
a) power cars?
b) make medicines?
c) power aeroplanes?
d) make roads?
e) power lorries?

2   An orange-flavoured fruit drop is bright orange in colour. Describe an experiment to find out if the fruit drop contains orange food colouring or a mixture of red and yellow food colourings.

3   Look at the diagram. It shows the simple apparatus you could use in the laboratory to separate the liquids in crude oil.
a) Explain how the liquid is transferred from the heated tube to the collecting test-tube.
b) Why is the collecting test-tube in a beaker of cold water?
c) Why is a thermometer used?
d) Why is the mineral wool used?
e) What safety measures would you take in this experiment?

Things to do

thermometer
delivery tube
crude oil soaked onto mineral wool
collecting test-tube
water
HEAT

Simple symbols

Often a symbol or picture gives information quicker than lots of writing.

▶ Test yourself on the following examples.
What do the symbols mean?

Elements

Chemists have a shorthand way of writing about elements.
They use **symbols** instead of writing out the names.

▶ Copy out the table. Fill in your guess for each element's symbol.
Then use the Help Sheet from your teacher to find out the correct answers.

Element	My guess for the symbol	Correct answer
carbon		
sulphur		
nitrogen		
oxygen		
fluorine		
phosphorus		

a Copy and complete the sentence to give a simple rule for writing the symbols for elements.
For some elements the symbol is the . . . . of the name of the element.

Now find the symbols for:

b calcium **c** chlorine **d** chromium.

The names of these elements all begin with the same letter.
The symbols use a second letter from the name too.
The second letter is written as a small letter.

e The symbols for copper, iron and sodium do not fit in with these rules. Where do you think their symbols come from?

Oxygen
Carbon
Magnesium
Calcium
Nickel
Boron
Germanium
Iodine
Cobalt
Lithium
Hydrogen

What **Ar**e **Th**e **S**ymbol **Ru**les? (Find the elements!)

Compounds

Remember that compounds form when 2 or more different atoms join together. The symbols for elements can be used to write a formula for a compound. For example,

CuO is copper oxide (ox**ide** when O is in a compound)
LiCl is lithium chloride (chlor**ide** when Cl is in a compound)

32

What do you think are the names for the following compounds?
Write them down.

f KCl   **g** CaO   **h** MgO   **i** NaCl

Some compounds have more complicated formulas.
Look at:

$CuCl_2$

This is copper chloride. The compound has 1 copper atom and 2 chlorine atoms. How can you tell this from the formula?

▶ Copy and complete the following table. The first one has been done for you.

Name	Formula	Number of each type of atom
carbon dioxide	CO_2	1 carbon, 2 oxygen
sodium fluoride		1 sodium, 1 fluorine
	$MgCl_2$	
	$AlCl_3$	
lithium oxide	Li_2O	

International science

All the symbols for elements are international. Maybe you can't understand the language. But you can spot the chemical elements! How's your Russian?

Use textbooks to find out about one of the elements mentioned in this Russian book.
Make a poster to show what you have found out.

ANALOGIJA MED DVODIMENZIONALNO RAZPOREDITVIJO IGRALNIH KART IN PERIODNIM SISTEMOM ELEMENTOV

Opredelitev problema:

Sredi 19. stoletja so kemiki opazili, da se kemijske lastnosti elementov periodično spreminjajo z naraščajočo (relativno) atomsko maso. Elemente so začeli razvrščati po kemijski sorodnosti. Tako je Wolfgang Döbereiner že leta 1829 razvrstil elemente po sorodnosti v trojke (tria-de).

Li	Cl
Na	Br
K	I

podobna tališča, gostote, podobna kemijska reaktivnost

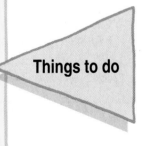

Things to do

1 Copy and complete the table:

Symbol	Name
C	
Cu	
	oxygen
N	
Ca	
	iron
	sodium
	chlorine
Mg	
S	

2 a) Write down the names of 2 elements in each case which are:
i) solids   ii) liquids   iii) gases.
b) Draw diagrams to show how the particles of the elements are arranged in solids, liquids and gases.

3 The table shows the approximate percentages of different elements in rocks of the Earth's crust.

Element	Percentage
oxygen	48
silicon	26
aluminium	8
iron	5
calcium	4
sodium	3
potassium	2
magnesium	2
other	2

Draw *either* a pie-chart *or* a bar-chart of this information.

Classifying elements

▶ Look at the objects in the photograph:
  Divide them into 2 groups:
  • Those you think are made from *elements*.
  • Those you think are made from *compounds*.

Materials can be sorted into groups. We say they can be **classified**.
You already know about some tests to classify materials.

▶ Look back to the Great Moon stones find on page 27.
  Which properties could you test?
  Make a list of the properties.

Some of these tests could be used to put *elements* into groups.

There are over 100 elements.
Some are hard to classify. Most can be put into 2 groups.
The groups are **metals** and **non-metals**.

▶ Copy out the table.
  Make it at least 10 cm long and 10 cm wide.
  Fill in your ideas about the properties of metals and non-metals.
  (You could check these with your teacher.)

Property	Metal	Non-metal
appearance		
strength		
hardness		
density		
melting and boiling points		
does it conduct heat?		
does it conduct electricity?		

Metal or non-metal?

Test the elements your teacher will give you.
Decide whether each element is a metal or non-metal.

The uses of an element depend on its properties.

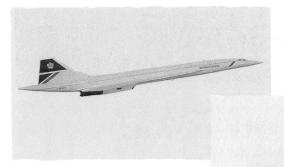

Why is aluminium used to make aircraft bodies?

Why is copper used to make saucepans?

Why is gold used to make jewellery?

Choose the cable

Liftum Ltd is a new company. It makes cables to carry cable cars. Each cable car seats 4 people.

Can you recommend a material to use for making the cables?
You could use copper, iron or aluminium.
Which would be best?

In your group, discuss which factors you will need to consider.
• Write out your list of factors.
• Which material do you *think* is the best?
• Why have you chosen this one? Are you happy with your choice?

Now **plan** an investigation on the 3 materials (copper, iron, aluminium).

What would it be useful to know about before recommending a material?

How could you find out about this?

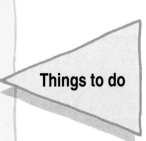

---

1 Write down the names and symbols of:
a) 5 metals   b) 5 non-metals.

2 Make a list of words which describe metals. Then make a list of words which describe non-metals.

3 Remi has found a lump of black solid. It is light, breaks easily and doesn't conduct electricity.
Is it a metal or non-metal?

4 Some objects can be made of metal or plastic. Discuss the advantages and disadvantages of metal or plastic for each of the following:
a) ruler   b) window frame
c) spoon   d) bucket.

5 Look around the room.
a) Name 4 objects made of metal.
b) Which metals are they made from?
c) Why were the metals used to make the objects?

Things to do

How reactive?

▶ Match each of the elements in the box with one of the descriptions **a** to **f**.

a an element with the symbol U
b a metal used to make cooking foil
c a metal used for electrical wires
d a metal which is a liquid
e an element with the symbol K
f a non-metal

- potassium
- uranium
- copper
- mercury
- carbon
- aluminium

You have tested properties of many materials. For example, strength and hardness. These are **physical properties**.

The **chemical properties** of a material are also important.
Does the material change easily into a new substance?
Does it **react?**

Remember! The use of a material depends on its properties.

Don't make a bridge from a metal which reacts with water!

Potassium is a metal. It reacts very violently with water. Potassium is stored under oil. Why do you think this is?

Reacting metals

Do metals react with oxygen in the air?

- Get a small piece of magnesium ribbon from your teacher. Hold it at arm's length in some tongs. Then move it into a Bunsen burner flame (air-hole just open). **Do not look directly at it**. What happens?

eye protection

- Do the experiment again using copper foil rather than magnesium. What happens?
  Which is the more reactive, magnesium or copper?

When metals react with oxygen they make new substances.
These are called **oxides**.

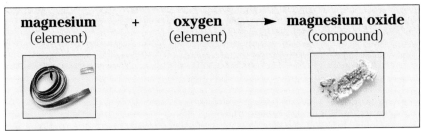

magnesium +	**oxygen** →	**magnesium oxide**
(element)	(element)	(compound)

Metals do not all react in the same way with oxygen or with water.

Some are *very reactive* – potassium.
Some are *reactive* – magnesium.
Some are *unreactive* – gold.

Metals can be put in an **order of reactivity**.
The most reactive ones are at the top of the list.
The least reactive are at the bottom.

Gold is unreactive. Why don't we use gold to make bridges?

What's the order?

Plan an investigation to produce an order of reactivity for metals.
You can use any equipment you need.
Remember to always use very small amounts of chemicals when investigating.

⚠ eye protection

The chemicals you can use are:
* *metal samples* – zinc, tin, magnesium, iron, copper,
* bottles of distilled (pure) water,
* bottles of dilute acid.

Order of Reactivity
Potassium
Magnesium
Gold

You **must** have your plan checked by your teacher.
Then do the investigation.
Write a report of your findings. Include your order of reactivity in your report.

1 Copy the diagram.
Put the correct words in the empty boxes.

metals compounds elements non-metals

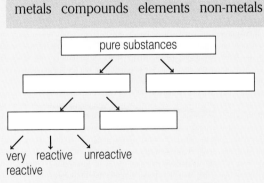

2 Use books to find out when different metals were discovered. Make a time-chart to display in the laboratory.
Look for a pattern between the discovery dates and the metals' reactivity.

3 Sue put some metals, A, B, C and D, in water. Look at the times taken for the metals to react completely with the water:

Metal	Time (seconds)
A	15
B	35
C	5
D	no reaction

a) Which is the most reactive metal?
b) Which metal could be copper?
c) Which metal is likely to be stored under oil?
d) How could Sue have made sure this was a fair test?

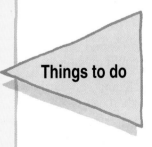

Things to do

Predicting reactions

You have seen that metals can be put in order of reactivity. This is called the **Reactivity Series**. It's a kind of League Table for metals. These tests are used to find the order.

| metal + air | metal + water | metal + acid |

The Reactivity Series

potassium	most reactive
sodium	
calcium	
magnesium	↑
zinc	
iron	
tin	
copper	least reactive

Can you make up your own rhymes to help you remember the order?
For example,
Please **S**top
Calling **M**y
Zebra **I**n
The **C**lass

► Use the Reactivity Series to help you answer these questions:

a You can put metals in acid. But your teacher will **not** give you samples of potassium, sodium or calcium for this. Why not?

b Where do you think gold fits in the order?

c Iron reacts slowly with water and air. What substance is made in this reaction? Do you have any examples of this at home?

d Copper does not react easily with air or water. Would it be a good idea to make cars from copper?

There is another way of finding out an order of reactivity for metals. We can set up **competitions**.
Competitions for oxygen are easy to do.

The big fight!

An experiment to heat magnesium oxide with copper is very boring. Nothing happens! There is no reaction.

Heating magnesium with copper oxide is much more exciting! There is a big reaction.

| **magnesium + copper oxide → magnesium oxide + copper** |
| (silver-grey) (black)   (grey-white)   (brown) |

This is because magnesium is more reactive than copper. Magnesium wins the fight for the oxygen.

Reactions like this are called **displacement reactions**. The magnesium **displaces** the copper. It pushes the copper out. It wins the oxygen.

Displacing metals

Try some other displacement experiments. See if you can spot reactions taking place. You should look to see if:
- a gas is made,

or
- any solids or solutions change colour,

or
- any solids disappear (dissolve) in solutions.

Take a spotting tray. Put **small** pieces of the 4 metals in the rows of the tray.
Use a teat pipette to add 4 different solutions to the 4 metals.
Check that yours looks like this:

Metals
• zinc
• iron
• magnesium
• copper

Solutions
• copper sulphate
• magnesium sulphate
• iron sulphate
• zinc sulphate

Pieces of: zinc iron magnesium copper

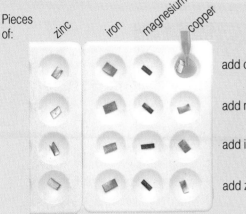

add copper sulphate solution

add magnesium sulphate solution

add iron sulphate solution

add zinc sulphate solution

Now you have added each solution to each metal. Have there been any reactions?

Yes ✓   ✗

Fill in the table with ticks or crosses:

	zinc	iron	magnesium	copper	
				✗	copper sulphate
			✗		magnesium sulphate
		✗			iron sulphate
	✗				zinc sulphate

e   Which of these rules is the correct one?
   i) Less reactive metals displace reactive ones.
   ii) Reactive metals displace less reactive ones.
   Copy out the correct rule.

---

1 Copy and complete:
In a reaction, a metal high in the Reactivity Series . . . . one below it. For example, ... could displace iron in a reaction.

2 Predict whether reactions will take place between these substances:
a) copper + zinc sulphate
b) iron + copper oxide
c) magnesium + iron nitrate
d) iron + potassium chloride
e) tin + magnesium oxide
f) zinc + copper oxide.

3 The metal nickel does not react with iron oxide. Nickel reacts with copper oxide.
a) Copy and complete:
   nickel + copper oxide → . . . . + ...
b) Explain why nickel won't react with iron oxide.
c) Alongside which metal would you put nickel in the Reactivity Series?

4 Carbon is an important non-metal. How could you put it in its right position in the Reactivity Series? What experiments could you do?

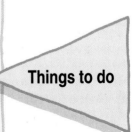

Things to do

Questions

1 What do you think each of the following words means?
Write no more than 2 lines for each.
a) atom b) molecule c) element d) compound e) mixture.

2 The 8 sentences below are about the process for getting pure salt from rock salt. Put the sentences in the right order so that the information makes sense.
Copy out the sentences when you have the right order.

- Rock salt is crushed into small pieces.
- The mixture is filtered.
- The solution is evaporated gently to leave dry salt.
- The pure salt dissolves.
- The mixture is warmed and stirred.
- The salt solution passes through the filter paper.
- Water is added to the mixture.
- The sand and dirt collect in the filter paper.

3 A sample of crude oil was found to contain the following:

Substance	Amount in crude oil (%)
refinery gas	0.2
gasoline	30.0
naphtha	7.0
kerosine	10.0
diesel oil	30.0
fuel oil	20.0
lubricating oil	2.0
bitumen	0.8

Put this information in the form of a bar-chart.

4 Metals can be mixed together to make **alloys**.
Find out the metals in each one of these alloys:
a) brass
b) solder
c) bronze
d) Duralumin.
Write down one use for each of the alloys.

5 Plan an investigation to put the following materials into an order of hardness:

   copper   iron   zinc   steel

6 In the Reactivity Series, carbon is usually placed just above iron. Carbon and copper oxide are both black powders. Unfortunately the labels have come off their containers in the laboratory.
What experiments could you do to find out which powder is which?

Food and Digestion

16

You need food to live.
Your food gives you your energy.
Without the right sort of food you won't grow strong
and healthy.

In this topic you will find out what happens to your food
when it is broken down inside your body.

Food for thought

Do you have any favourite foods?

▶ Make a list of some of the foods that you like to eat.

Which of these foods do you think are good for you?

Draw a ring around those foods that are not good for you.

Healthy eating

You need a healthy diet to:
- grow   • repair damaged cells   • get energy   • keep healthy.

A healthy diet should include some of each of these:

Proteins are for growth. They are used to make new cells and repair damaged tissue.

Carbohydrates, like sugar and starch, are our high-energy foods, but eat too much and they turn to fat.

Fats are used to store energy. They also insulate our bodies so that we do not lose a lot of heat.

Vitamins and minerals are needed in small amounts to keep us healthy e.g. iron for the blood and vitamin D for the bones.

The easiest way to have a healthy diet is to eat a variety of foods each day.

▶ Choose one food from each picture to plan some healthy meals.

▶ Look at the newspaper report:

a What is meant by 'junk food' or 'fast food'?

b Why do you think a lot of children eat junk or fast food
i) at home? ii) in school?

c Why is it important to eat fresh fruit and vegetables?

Survey shows junk food is favourite

Kids ditch salad for the chips

JUNK food is children's first love at lunchtime, a survey disclosed today.

Pizzas, burgers and hot dogs are top of the menu for school dinners, with pupils giving scarcely a second thought to healthy eating.

Fresh fruit is losing out to junk food. Children eat only 60% of the fruit they did 4 years ago.

Food tests

Here are 4 ways of testing for foods.

Do each test carefully and observe the result.

Write down your results in each case.

Testing for starch

Add 2 drops of *iodine* solution to some starch solution.

What do you see?

Testing for glucose

Add 10 drops of **Benedict's solution** to one-third of a test-tube of glucose solution. Heat carefully in a water bath.

What do you see?

eye protection

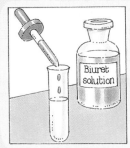

Testing for protein

Add 10 drops of **Biuret solution** (be careful: this is corrosive) to half a test-tube of protein solution.

What do you see?

⚠ eye protection

Testing for fat

Rub some of the food onto a piece of filter paper.

Hold the paper up to the light.

What do you see?

Now try testing a few foods. If the food is a solid you will have to grind it up with a little water first.

Record your results in a table like this:

Food	Starch	Glucose	Protein	Fat
Nuts			✓	✓

Things to do

1 Copy and complete the table:

Food	Use to my body	Food containing a lot of it	Chemical test
protein starch fat glucose			

2 Do some research to find out what these vitamins and minerals are needed for. What happens if you do not get enough of them?
a) vitamin C
b) iron
c) vitamin A
d) calcium
e) vitamin B group
f) iodine.

3 Keep a careful record of all the food that you eat in the next 24 hours.
Use the Recommended Daily Amounts table to see if you had a healthy diet.

4 Make a survey of eating habits of the people in your class. Look for any patterns in the results. For instance, how much fatty food, junk food or fibre do they eat?

Teeth and decay

Do you have healthy teeth?
They make your mouth look good and feel good.
Your teeth are important and you should remember to take care of them.
They chew your food up into small pieces before you swallow it.
Imagine trying to swallow an apple whole!

a　Which foods could you eat if you had no teeth?

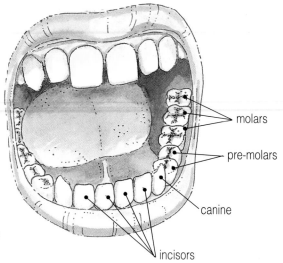

molars

pre-molars

canine

incisors

Types of teeth

▶ Look at your teeth in a mirror.

When you are an adult you should have 32, but you won't have all of them yet!

Look carefully for the 4 types of teeth.

b　Write down what you think each of them is used for.

▶ Your teacher will give you a diagram of a set of teeth.

Shade in the ones in which you have fillings and cross out any that are missing.

Parts of a tooth

The part of a tooth that you can see in your mouth is covered with a white layer of **enamel**.
Enamel is the hardest substance in your body.

c　Why is it found here?

enamel

dentine

pulp cavity

gum

cement

nerves and blood vessels

Underneath is the living part of your tooth which is made of **dentine**. This is softer than enamel.

In the middle of each tooth is the **pulp cavity** which contains nerves and blood vessels.

d　When you have a tooth out, why does it bleed and feel sore?

Do you brush your teeth properly?
▶ Find out about how you should brush your teeth.
Your teacher may give you a Help Sheet.

Remember, you can keep your teeth and gums healthy by:

CLEAN YOUR TEETH at least TWICE A DAY
after breakfast and before going to bed!

REPLACE YOUR TOOTHBRUSH
when it wears out
every 4-6 months

AVOID SUGARY FOOD & DRINK
Between meals

VISIT YOUR DENTIST REGULARLY EVERY 6 MONTHS

Under attack

Tooth decay can spoil your looks and cause pain.
It is the most common disease affecting school children in Britain.
Before brushing, do your teeth feel rough and sticky in places?
This is **plaque**. It can form when food sticks to the surface of your teeth.

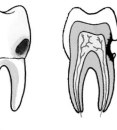

1 Bacteria grow on the food (especially sugary food) and form plaque.

3 Food collects in the cavity and bacteria make more acid. This acid attacks the enamel and the cavity gets bigger.

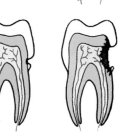

2 The bacteria make acid. This acid attacks the enamel and makes a small hole (a **cavity**).

4 Once through the enamel the cavity quickly spreads through the dentine to the pulp. The nerve is now affected. It can be very painful.

Plaque attack

How effective are different toothpastes at killing bacteria?

1 Take an agar plate with harmless bacteria growing on it. Holes have been cut out for you in the jelly.
2 Using a marker pen, number each hole on the underside of your plate.
3 Then lift the lid and carefully fill each hole with a different toothpaste, making a note of what is in each hole.
4 Replace the lid and fix it down with adhesive tape.
5 Place your agar plate in a warm incubator at 25 °C for 2 days.
6 Wash your hands with soap and water.
7 After 2 days, measure the diameter of any clear areas.
Sketch your results on a copy of the diagram.
Do not open the plate.
When you have finished, return the plate to your teacher.

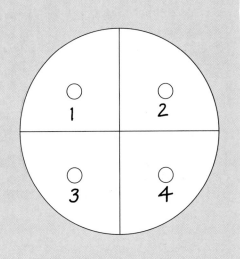

e What effects do the different toothpastes have on the bacteria?
f Which toothpaste do you think is best to use and why?
g Many toothpastes are alkaline. How does this help them fight tooth decay?

1 Copy and complete:
The hard coating on the outside of a tooth is called . . . . It surrounds a softer layer called . . . . In the middle of the tooth is the pulp . . . . which contains . . . . vessels and . . . .

2 If fluoride is added, in small amounts, to drinking water, it can reduce tooth decay. In large amounts fluoride can be poisonous to humans. Carry out a survey to find out what people think about this problem.

3 Take half a raw carrot and use it to find out what each type of tooth does. Nibble or gnaw off a small piece, then bite off a larger one. Tear off a piece like a dog tears meat. Finally chew it and grind it up for swallowing. Record how each of your teeth is used.

4 Suppose you have to give a 2-minute talk to some junior school children about caring for their teeth. Write down what you would say.

Things to do

Digestion

Think of the different foods that you eat.
How much of your food is **soluble** (will dissolve in water)?
Probably not much.

Before our bodies can use the food that we eat it must be **digested**.
When food is digested it is broken down into very small molecules:

There are special **digestive juices** in our body.
These digest large molecules into small ones.

▶ Try chewing some bread for a long time. Eventually it tastes sweet
because your saliva has broken down the starch in the bread to sugar.

Starch is a very big molecule. It is made up
of lots of sugar molecules joined together.

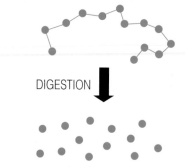

DIGESTION

sugar molecules are very small

Changing starch into sugar

You can find out how saliva affects starch by carrying out this
experiment.

1 Set the 2 test-tubes up as shown in the diagram.
2 Leave the apparatus for 10 minutes at 40 °C.
3 Test a drop from each test-tube for starch with iodine. What do
   you see?
4 Add some Benedict's solution to each test-tube and test for
   sugar. What do you see?
5 Record your results in a table like this:

	Colour with iodine	Colour with Benedict's
test-tube A test-tube B		

a In which test-tube was the starch broken down?
b What do you think the starch was broken down to?
c What do you think boiling did to the saliva?
d Why were your test-tubes kept at 40 °C?

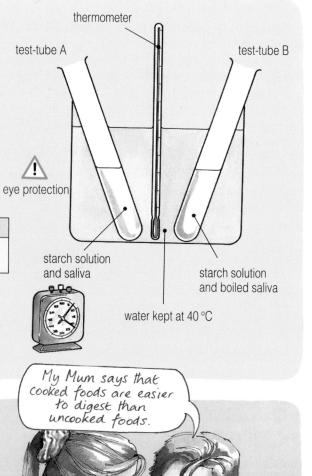

thermometer

test-tube A   test-tube B

⚠
eye protection

starch solution
and saliva

starch solution
and boiled saliva

water kept at 40 °C

What's cooking?

Sam and Charlotte were talking about digesting food:

Plan an investigation to see who is right.
You could use some starchy foods like pasta or potato.
Remember that you must make it a fair test.
When you have decided upon your plan, show it to your teacher,
then try it.

My Mum says that
cooked foods are easier
to digest than
uncooked foods.

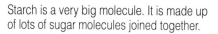

I think you are wrong.
I think it is easier to
digest uncooked foods.

How is starch digested to sugar?

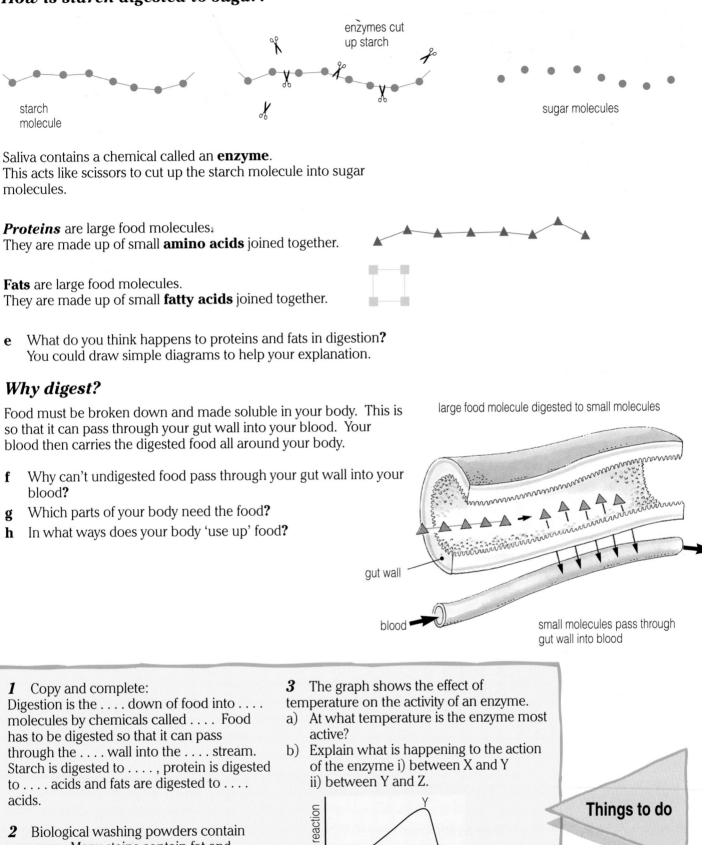

enzymes cut
up starch

starch
molecule

sugar molecules

Saliva contains a chemical called an **enzyme**.
This acts like scissors to cut up the starch molecule into sugar
molecules.

Proteins are large food molecules.
They are made up of small **amino acids** joined together.

Fats are large food molecules.
They are made up of small **fatty acids** joined together.

e What do you think happens to proteins and fats in digestion?
You could draw simple diagrams to help your explanation.

Why digest?

Food must be broken down and made soluble in your body. This is
so that it can pass through your gut wall into your blood. Your
blood then carries the digested food all around your body.

large food molecule digested to small molecules

f Why can't undigested food pass through your gut wall into your
blood?
g Which parts of your body need the food?
h In what ways does your body 'use up' food?

gut wall

blood

small molecules pass through
gut wall into blood

---

1 Copy and complete:
Digestion is the . . . . down of food into . . . .
molecules by chemicals called . . . . Food
has to be digested so that it can pass
through the . . . . wall into the . . . . stream.
Starch is digested to . . . . , protein is digested
to . . . . acids and fats are digested to . . . .
acids.

2 Biological washing powders contain
enzymes. Many stains contain fat and
protein that can be digested by enzymes.
Plan an investigation into the effect of
temperature on biological and non-
biological washing powders.

3 The graph shows the effect of
temperature on the activity of an enzyme.
a) At what temperature is the enzyme most
active?
b) Explain what is happening to the action
of the enzyme i) between X and Y
ii) between Y and Z.

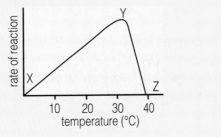

rate of reaction

10   20   30   40
temperature (°C)

Things to do

47

16d *Your gut*

What happens to your food when you swallow?
It enters a tube that starts with your mouth and ends at your anus.
The whole of this food tube is called your **gut**.

Your gut is about 9 metres long.

▶ Work out how many times your height it is.

a How does all this length of gut fit into your body?

b Why do you think your gut has to be so long?

Look at the diagram of the human gut below.

▶ Follow the path your food goes down. There are lots of twists and turns.

c Write down the correct order of parts that food passes through.

Down the tube

Mouth
Food chewed and mixed with saliva. Then you swallow it (gulp!).
(Food is here for 20 seconds)

Gullet
A straight, muscular tube leading to your stomach.
(10 seconds)

Stomach
The acid bath! Digestive juices and acid are added to food here. Your stomach churns up this mixture.
(2 to 6 hours)

Small intestine
More juices are added from your liver and your pancreas. These complete digestion. Then the food passes through into your blood.
(About 5 hours)

Large intestine
Only food that can not be digested (like fibre) reaches here. A lot of water passes back into your body. This leaves solid waste to pass through your anus.
(Up to 24 hours)

d In which part of your gut does food stay the longest? Why do you think this is?

e Proteins are digested in your stomach. What are conditions like here?

f How long does it take food to pass down the whole length of your gut?

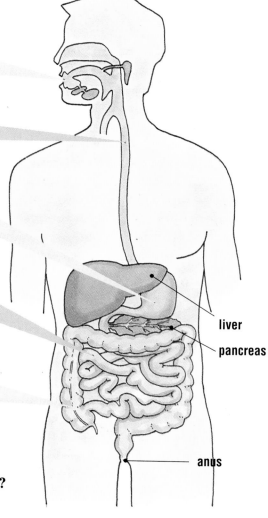

liver

pancreas

anus

A model gut

You can make a model gut using **Visking tubing**.

1 Wash the Visking tubing in warm water to soften it.
2 Tie one end in a tight knot.
3 Use a syringe to fill the tubing with 5 cm of starch solution and 5 cm of glucose solution.
4 Wash the outside of the tubing.
5 Support your model gut in a boiling tube with an elastic band.
6 Fill the boiling tube with water and leave for 15 minutes.
7 After 15 minutes, test the water for starch and for sugar.

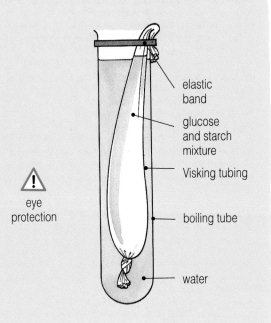

⚠ eye protection

elastic band

glucose and starch mixture

Visking tubing

boiling tube

water

g Which food passed through the tubing into the water? How do you explain this?

h Which food did not pass through the tubing into the water? How do you explain this?

i Which part of the apparatus was like: i) food in your gut? ii) your gut wall? iii) the blood around your gut?

How does food pass along your gut?

Muscles in your gut wall squeeze your food along.
It's like squeezing toothpaste out of a tube.
But muscles need something to push against and that's where **fibre** is helpful.
Fibre cannot be digested so it isn't broken down.
Fibre adds bulk and a solid shape to your food so that it can be pushed along your gut.
Tough, stringy plants like celery have lots of fibre.

▶ Make a list of foods that are high in fibre.

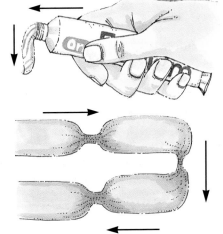

---

1 Match the parts of the body in the first column with the descriptions in the second column:

a) stomach
b) small intestine
c) large intestine
d) mouth
e) gullet

i) most water is absorbed here.
ii) saliva is made here.
iii) most food is absorbed here.
iv) carries food down to the stomach.
v) is very acidic.

2 Find out how each of the following parts of the body help digestion to take place:
a) liver   b) pancreas   c) appendix.

3 The following are diseases of the gut:
a) constipation   b) stomach ulcers
c) diarrhoea.
Find out the causes of each of these diseases.

Things to do

Questions

1 Do some research to find a diagram of the teeth of a carnivore e.g. a cat or a dog. Copy the diagram and write down ways in which the teeth are adapted for eating meat.
Do the same for the teeth of a herbivore e.g. a rabbit or sheep. How are they adapted for eating plants?

2 When some foods are made, chemicals are added. We call these chemicals **food additives**. Some of them can make the food last longer. Others can give the food a better flavour or a better appearance. Carry out a survey into the food additives found in your kitchen. Look at the food labels and then list the additives. Some may have a chemical name like monosodium glutamate or an 'E number' like E330.
Try to find out why they have been put in the food.

3 Plan an investigation to compare the amount of water in a piece of plant food with the amount of water in a piece of meat. You can use the sort of apparatus found in your science laboratory. Remember to make it a fair test and check your plan with your teacher before carrying it out.

4 Visit your local supermarket. Find out the cost of different food groups using the examples in the table:
   a) Which food can you buy most of for £5?
   b) Which food can you buy least of for £5?

Some families have more to spend on food than others.
   c) Which food groups do you think a family on a very low income would have to buy, to feed themselves?
   d) Which food groups would a high-income family be able to buy?
   e) Do you think your answers to c) and d) would be true in India and in the United States?

Food group	Example	Cost per 1 kg
carbohydrates	potatoes rice	
fats	cheese butter	
proteins	chicken lamb	
vitamins and minerals	oranges broccoli	

5 Your teacher will give you a table of Recommended Daily Amounts of nutrients (RDAs).
   a) Write down how much energy you need. How does this compare with:
      i) a 1 year old?
      ii) someone the same age as you but of the opposite sex?
      iii) a very active adult female?
   b) Which group shows the biggest increase in protein needs for each sex?
      Why do you think this is?
   c) Which foods does a pregnant woman need more of than a female desk worker? Try to explain these differences.

6 Plan the following meals, choosing foods that would be good for you:
   a) A good breakfast for a 12 year old.
   b) A high-energy lunch for an athlete before a big race.
   c) An evening meal low in fat but rich in fibre and protein.

Earth and Space

17

Here is a photo of our beautiful planet Earth.
It is one of the 9 planets that go round the Sun.

Our Sun is just one of the billions of stars in our galaxy.

And our galaxy is just one of the billions of galaxies in
our Universe . . . .

Each morning, the Sun rises in the East.

a In which direction does it set at dusk?

b In which direction is it at mid-day?

c Why must you never look straight at the Sun?

d In winter, is the day-time shorter or longer than in summer?

e In winter, is the Sun higher or lower in the sky?

f What would happen on Earth if the Sun stopped shining?

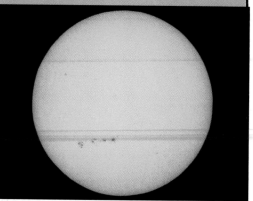

The Sun – our nearest star. On the same scale, the Earth is about the size of this full stop .

Day and night

Use a ball and a lamp (or a torch) to find out why we get day and night:

Sun

Earth

g If it is day-time for you, name a country where it is night-time.

h How many hours does it take for the Earth to spin round once?

i Which way does the Earth spin so that the Sun 'rises' in the East?

A year

Use a ball and a lamp to find out how the Earth moves in orbit round the Sun:

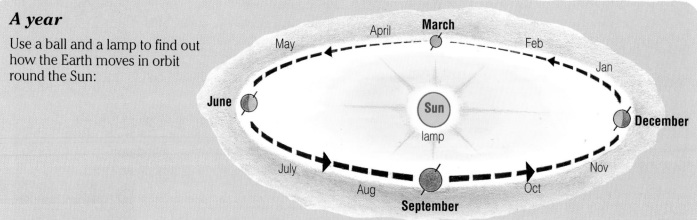

j How long does it take for the Earth to make one complete journey round the Sun?

k How many times does the Earth spin on its axis while it makes this journey?

l You are held on to the Earth by the force of gravity. What force do you think keeps the Earth in orbit round the Sun?

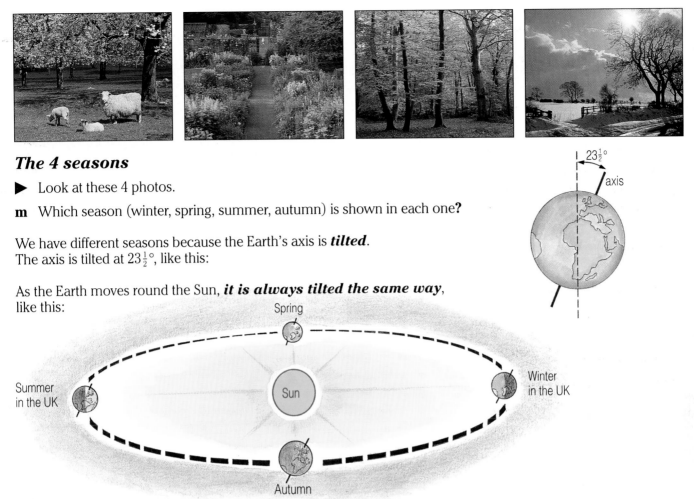

The 4 seasons

▶ Look at these 4 photos.

m Which season (winter, spring, summer, autumn) is shown in each one?

We have different seasons because the Earth's axis is **tilted**.
The axis is tilted at $23\frac{1}{2}°$, like this:

As the Earth moves round the Sun, **it is always tilted the same way**, like this:

In summer, our part of the Earth is tilted towards the Sun. The Sun appears to be higher in the sky, and daylight lasts longer. So it is warmer.
In winter, our part of the Earth is tilted away from the Sun. The Sun is lower in the sky, and the day-time is shorter. So it is colder.

Use the ball and lamp to show the 4 seasons. Mark your position on the ball, and watch it carefully as the Earth goes round the Sun. (Remember to keep the ball tilted in the same direction all the time.)

▶ Imagine you are at the North pole. At what time of the year is it

n daylight for 24 hours?

o night-time for 24 hours?

1 Copy and complete:
a) A day is the time for the . . . . to . . . . once on its axis.
b) A year is the time it takes for the . . . . to travel once round the . . . .
c) In one year there are . . . . days.
d) The Earth's axis is tilted at an angle of . . . .
e) In summer, our part of the . . . . is tilted towards the . . . . , so the Sun appears . . . . in the sky and the days are . . . . and warmer.

2 A scarecrow, 1 metre high, is standing in the middle of a field. Write down as many things as you can about its shadow,
a) in summer,   b) in winter.

3 How would our lives be different if:
a) The Earth was much closer to the Sun?
b) The Earth turned more slowly on its axis?
c) The Earth's axis was not tilted at all?

Things to do

The Earth and the Moon

▶ Look at the photos:

a Write down 5 things that you know about the Moon.

b Would you like to live on the Moon? Why?

c The Moon shines at night, but it is not hot like the Sun. Where do you think the light comes from?

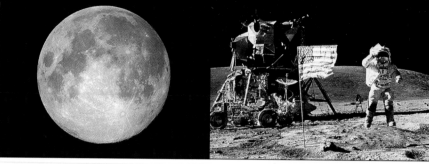

Full Moon   An astronaut on the Moon

The Moon moves in an orbit round the Earth.
It is held in this orbit by the pull of gravity.
One complete orbit of the Moon takes about 1 month (1 'moonth').

The Moon looks different at different times of the month.
It has **phases**. A 'full moon' is one of the phases.

Phases of the Moon

Use a lamp and 2 balls to investigate the phases of the Moon:

The numbers 1–8 show 8 different positions of the Moon round the Earth. They are about 4 days apart.

At each position, look at the Moon from the position of the Earth. That is, from the **centre** of the circle.

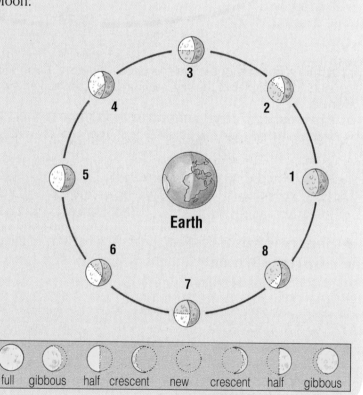

On this diagram, some parts of the Moon are coloured yellow. These are the parts in sunlight that you can see from the Earth.

• Sketch what you see in each position when you are at the centre of the circle. Label your sketches with the correct names of the phases:

full   gibbous   half   crescent   new   crescent   half   gibbous

Observing the Moon

Your teacher will give you a Help Sheet on which you can record your observations of the Moon for the next month.

Eclipse of the Moon *(lunar eclipse)*

When you stand in sunlight, there is a shadow behind you.
In the same way, there is a big shadow behind the Earth.
If the Moon moves into this shadow, we call it an **eclipse** of the Moon:

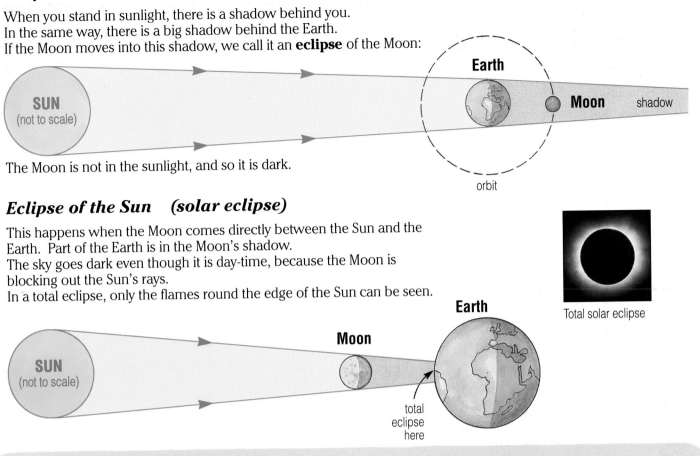

The Moon is not in the sunlight, and so it is dark.

Eclipse of the Sun *(solar eclipse)*

This happens when the Moon comes directly between the Sun and the
Earth. Part of the Earth is in the Moon's shadow.
The sky goes dark even though it is day-time, because the Moon is
blocking out the Sun's rays.
In a total eclipse, only the flames round the edge of the Sun can be seen.

Total solar eclipse

Use a lamp and 2 balls to show:
1) an eclipse of the Moon, and   2) an eclipse of the Sun.

The Moon is covered in craters. We think they were caused by
large rocks from space, crashing into the Moon.
These rocks are called **meteorites**.

Design an investigation to find out what changes the *size and
shape of craters*. (Hint: you could use sand and marbles.)

Plan the investigation, and if you have time, do it.

Things to do

1   Copy and complete:
a) The Moon takes one . . . . to go round
    the . . . . In each position it looks
    different to us, with different . . . .
b) In an eclipse of the Moon, the Moon
    moves into the shadow of the . . . .
c) In an eclipse of the Sun, the . . . . blocks
    out the light from the . . . . so that the
    . . . . is in a shadow.

2   Design a Moon-station for an astronaut
to live in. Draw a plan and label all the
important features.

3   Draw a diagram of the Earth and Moon
to a scale of: 1 mm = 1000 miles.

Earth–Moon distance	= 240 000 miles
Earth's diameter	= 8000 miles
Moon's diameter	= 2000 miles

The Sun is 93 000 000 miles away, and
900 000 miles in diameter. Where would
the Sun be on your diagram?

4   Write a story about a voyage to the
Moon. Describe some of the difficulties
you would have to overcome.

The Earth and other planets

The Earth is a **planet**. It travels in an orbit round our star, the Sun.

a Which is bigger: the Sun or the Earth?
b How long does it take for the Earth to make 1 orbit of the Sun?

The Earth is one of a 'family' of 9 planets. All of them are orbiting round the Sun. This is the **Solar System**.

The 9 planets are different sizes:

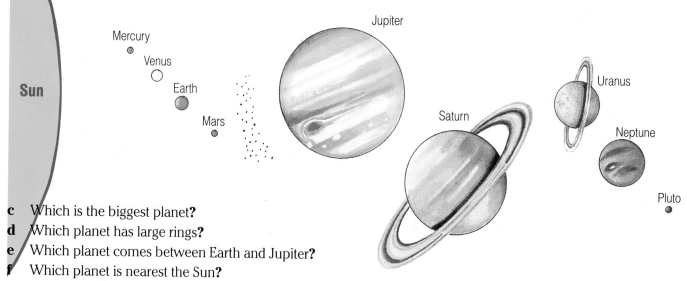

c Which is the biggest planet?
d Which planet has large rings?
e Which planet comes between Earth and Jupiter?
f Which planet is nearest the Sun?
g Which planet is farthest from the Sun?
h Which planet do you think will be the coldest?

Here are some data on the planets:

	Mercury	Venus	Earth	Mars	Asteroids	Jupiter	Saturn	Uranus	Neptune	Pluto
Diameter (km)	5000	12 000	12 800	7000	–	140 000	120 000	52 000	50 000	3000
Distance from the Sun (million km)	60	110	150	230	–	780	1400	2900	4500	6000
Time to travel 1 orbit round the Sun (years)	0.2	0.6	1	2	–	12	30	84	160	250

i Which planet is almost the same size as the Earth?
j Which planets are larger than the Earth?
k Which planet moves round the Sun in the shortest time?
l What pattern can you see between the *distance* from the Sun and the *time* taken for 1 orbit?
m What are the asteroids?

How far apart are the planets?

The distances between the planets are huge – much farther than the diagram on the opposite page shows.

Here is a scale diagram of the distances to the planets:

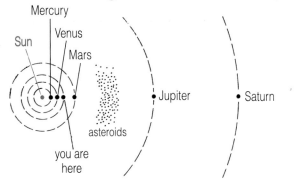

n Pluto is not shown on this diagram. Where would it be?

o Write down the names of the 4 inner planets.

p Why are these inner planets hotter than the 5 outer planets?

q Would the Sun look bigger or smaller from Mercury?

r Would the Sun look bright or dim from Pluto?

s What is the name of the force that holds the planets in orbit round the Sun?

t The orbit of each planet is not quite a circle. It is an **ellipse**. Draw an ellipse.

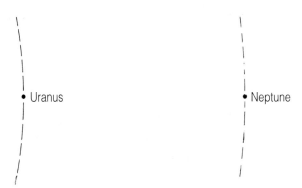

The Voyager-2 space probe

Make a scale model of the Solar System

1 For the Sun use a grapefruit or a cardboard disc with a diameter of 11 cm.

2 For the Earth make a small ball of plasticine just 1 mm across.
Make all the other planets to the same scale, using the table below:

3 Hold your 'Earth' at a distance of 12 metres from your 'Sun'. Use the table to hold all the other planets at the correct distances.
You will need to go on the playing field!

On this scale the nearest star would be another grapefruit, about 3000 kilometres away!

	Mercury	Venus	Earth	Mars	Asteroids	Jupiter	Saturn	Uranus	Neptune	Pluto
Size of 'planet'	$\frac{1}{2}$ mm	1 mm	1 mm	$\frac{1}{2}$ mm		11 mm	9 mm	4 mm	4 mm	$\frac{1}{4}$ mm
Distance from 'Sun'	5 m	8 m	12 m	18 m		60 m	110 m	220 m	350 m	460 m

1 Copy and complete:
a) There are .... planets in the .... System.
b) The names of the 9 planets (in order) are:
c) The coldest planet is .... This is because it is the farthest from the ....

2 Why do you think Pluto was the last planet to be discovered?

3 What do you think it would be like to live on Mercury?

4 Plot a bar-chart of the diameters of the planets.

5 For the first 5 planets, plot a line-graph of the *time* taken to travel 1 orbit round the Sun against the *distance* from the Sun.

The asteroids are large rocks that travel round the Sun at an average distance of 400 million km. Use your graph to estimate how long they take to make 1 orbit.

Things to do

The Solar System

Use the information on these two pages to fill in a table like this one:

Planet	Type of surface	Average temperature	Type of atmosphere	Length of a 'day'	Moons, rings
Mercury					

Mercury is a small planet, about the size of our Moon. It has a rocky surface which is covered in craters.

It has no atmosphere. The side facing the Sun is very hot (about 430 °C, hot enough to melt lead).

Venus is almost as big as the Earth, but it is very unpleasant. Its rocky surface is covered by thick clouds of sulphuric acid.

The atmosphere is mainly carbon dioxide. This traps the Sun's heat (by the 'Greenhouse Effect') so that Venus is even hotter than Mercury.

From space, **Earth** is a blue planet with swirls of cloud. It is the only planet with water and oxygen and living things.

It is at the right distance from the Sun, with the right chemicals, to support life. Of course, other stars in the Universe may have planets with the same conditions.

Mars – the red planet – is a dry cold desert of red rocks, with huge mountains and canyons. There is no life on Mars.

It has a thin atmosphere of carbon dioxide, and 2 small moons. Mars was the first of the planets to be visited by one of our space-craft.

Planet	Diameter (km)	Distance to Sun (million km)	Time for 1 orbit (planet's 'year')	Time for 1 spin (planet's 'day')	Average temperature on sunny side (°C)	Moons
Mercury	5000	60	88 days	1400 hours	+430	0
Venus	12 000	110	220 days	5800 hours	+470	0
Earth	12 800	150	$365\frac{1}{4}$ days	24 hours	+20	1
Mars	7000	230	2 years	25 hours	−20	2
Asteroids						
Jupiter	140 000	780	12 years	10 hours	−150	16
Saturn	120 000	1400	30 years	10 hours	−180	18 + rings
Uranus	52 000	2900	84 years	17 hours	−210	15 + rings
Neptune	50 000	4500	160 years	16 hours	−220	8
Pluto	3000	6000	250 years	150 hours	−230	1

Jupiter is the largest planet, and is very cold. It has no solid surface. It is mainly liquid hydrogen and helium, surrounded by these gases and clouds. The Giant Red Spot is a huge storm, 3 times the size of Earth. Jupiter has 16 moons.

Saturn is another 'gas giant', very like Jupiter. The beautiful rings are not solid. They are made of billions of chunks of ice and rock. They are held in orbit by the pull of Saturn's gravity.

Uranus is another 'gas giant', made of hydrogen and helium.
Unlike the other planets it is lying on its side as it goes round the Sun.
It was discovered in 1781 by William Herschel.

Neptune is very like Uranus. It is a blue 'gas giant'. The Great Dark Spot is a storm the size of Earth.

a Which planet is most like the Earth? Explain your reasons.
b Why is it hard for scientists to find out about i) Venus? ii) Pluto?
c Only one planet has liquid water on its surface. Why is this?

Pluto is the smallest planet, discovered in 1930.
It is a rocky planet, covered in ice. It has a very thin atmosphere of methane.

America is planning to send astronauts to Mars.
It would cost billions of dollars. Do you think it is worth it?

Discuss this in your group, and write down the arguments for and against.

1 Imagine that you are an advertising agent for holidays in the year 2020. Choose one of the planets (not Earth) and:
a) make up an advertising slogan for it,
b) draw a poster or write a TV commercial for it.

2 Write a story about 'A journey through the Solar System'.

3 Explain why you think life developed on Earth and not on other planets.

4 Which planets have a thin atmosphere or none at all?
Use your data to see if it has anything to do with size.
Can you think of a reason for this?

Things to do

Our place in the Universe

Our Sun is a **star**. It is like all the others you can see in the night sky. In size, the Sun is just an average star.

The star patterns you can see at night are called **constellations**.
For example, the Plough (or Great Bear) looks like this:

► Write down the names of any constellations that you know of. Your teacher may give you a star-map of the constellations.

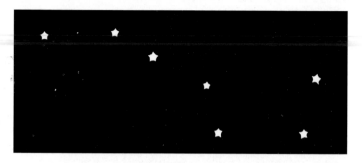

The Sun is part of a huge collection of stars called a **galaxy**. Our galaxy is called the Milky Way. It is a collection of more than 100 000 million stars!

Our galaxy has a *spiral* shape. We are in one of the spiral arms:

Our galaxy is huge. It takes light just 8 minutes to travel from the Sun to Earth, but it takes 100 000 years for light to travel across our galaxy!

A **light-year** is the *distance* that light travels in one year. And light travels at a speed of 300 000 kilometres per second!

Radio waves also travel at the speed of light. Nothing can travel faster than this.

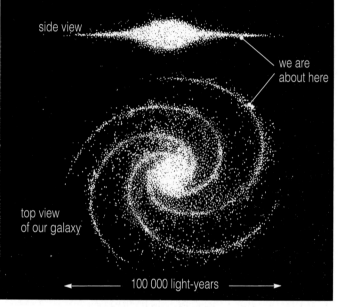

side view

we are about here

top view of our galaxy

100 000 light-years

Our galaxy, the Milky Way

Other galaxies

The Milky Way is our galaxy, but it is not the only galaxy. It is one of a group of 20 galaxies called the **Local Group**.
The Andromeda galaxy is one of these:

Through telescopes we can see *millions* of other galaxies!
All the galaxies together, and the space between them, form the **Universe**.

Some galaxies are so far away that it has taken the light 10 000 million years to reach us. So we see them as they *were*, 10 000 million years ago!

So the universe is even older than this.

The Andromeda galaxy.
It contains 300 billion stars and is
2 million light-years away from us.

The expanding universe

In 1929, Edwin Hubble discovered that the galaxies are moving apart. The universe is expanding!

This is rather like a balloon which has some ink-marks on it. The ink-marks represent the galaxies. The balloon is the universe. As the balloon is blown up, the universe expands and all the galaxies move farther apart.
This is a 'model' of our expanding universe.

Thinking back in time, the universe was once very small. Astronomers believe it started about 15 000 million years ago, in an explosion called the **Big Bang**.
It has been expanding ever since.

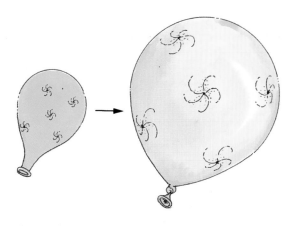

Your place in the universe

Your teacher will give you a Help Sheet for this.
Cut out the pictures and sort them into the right order.
This will show you how you fit into the universe.

* What is your full address in the universe?

Making a telescope

To look at the universe, an astronomer uses a **telescope**.

You can make a telescope by using 2 lenses, like this:

Look through the lenses, and move the thin lens along the ruler until you see a sharp image.

* What do you notice about the image that you see?

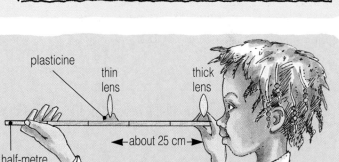

ET ... Extra-Terrestrial

Do you think there could be other life in the universe? Perhaps on a planet round another star?

Suppose you were going to send a 'space-capsule' on a long journey into space. It may be found by aliens some time in the future.

What would you put into the capsule to tell the aliens about yourself? (Remember: they won't understand English!)

---

1 Copy and complete:
a) The Sun is really an ordinary . . . . It is part of our . . . . , called the Milky Way.
b) A light-year is the . . . . that light travels in one . . . .
c) The universe has been . . . . since the time of the . . . .

2 The speed of light is 300 000 km/s. How far, in kilometres, is a light-year?

3 Here is a list of objects:
star moon galaxy planet universe
a) Put them in order of size (smallest first).
b) For each one, write a sentence to explain what it is.

4 To travel to another star would take centuries. Sketch the design of a space-ship for this. What would be the problems for the people on board?

Things to do

1 The table shows some data for the sunshine in London:

a) Explain why summer is hotter than winter.
b) Sketch the path of the Sun through the sky for
    i) a day in January, and   ii) a day in July.

Month	Altitude of midday sun	Hours of daylight
January	low, 15°	8
July	high, 62°	16

2 The table shows the times of sunrise and sunset in London throughout the year:

Date	Jan 21	Feb 21	Mar 21	Apr 21	May 21	Jun 21	Jul 21	Aug 21	Sep 21	Oct 21	Nov 21	Dec 21
Sunrise	8.0	7.2	6.0	5.0	4.1	3.7	4.1	4.8	5.8	6.6	7.4	8.1
Sunset	16.3	17.3	18.2	19.1	19.8	20.3	20.1	19.3	18.1	17.0	16.1	16.0

(All the times are in decimal hours and GMT on a 24-hour clock)

a) On graph paper, plot the sunrise times against the date.
   Then plot the sunset times on the same diagram.
b) When is the day longest?
c) When is the day shortest?
d) When is the day-time equal in length to the night?

3 The photo shows the first astronaut to land on the Moon:

a) Describe first of all what you can see in the photo.
   These are your **observations**.

b) Then write down what **conclusions** you can make from:
   i)   his clothes,
   ii) his shadow,
   iii) the black sky,
   iv) his foot-marks,
   v)   his small space-craft,
   vi) the label under the photograph.

Neil Armstrong, 1969

4 Some pupils are thinking and hypothesizing:

Ayesha says, "I think that the farther a planet is from the Sun, the cooler it is."
Danielle says, "I think that the larger a planet is, the more moons it has."
Chris says, "I think that the farther a planet is from the Sun, the longer the time for
1 orbit (the planet's 'year')."

Do the data in the table on page 58 support any of these hypotheses? Explain your
thinking. If you can, draw graphs to show how the data agree with the hypotheses.

5 Using an astronomy book or an encyclopedia, write a paragraph about each of these:
a) a supernova,   b) a neutron star (pulsar),   c) a black hole,   d) a quasar.

6 In 1670, Blaise Pascal, a French scientist, wrote "Le silence éternel de ces espaces
infinis m'effraie" (*The eternal silence of these infinite spaces terrifies me*). Write a
poem of your thoughts about space.

Staying alive 18

How long could you live without oxygen? Not long.
Your lungs take oxygen out of the air.
In your lungs the oxygen passes into your blood.
Your heart pumps the blood all round your body.
No wonder your heart and lungs are so important!

In this topic:
- **18a** *Breathing*
- **18b** *A change of air*
- **18c** *Dying for a smoke?*
- **18d** *Living liquid*
- **18e** *Heartbeat*
- **18f** *Liquid protection*

Breathing

We all exercise at some time.
What exercise have you had in the last week?

Do you feel different after exercise?

▶ Write down some of the changes that happen to your body when you exercise.

You can find out how exercise affects you by doing this investigation:

Puffing and panting

1 Sit still and count how many times you breathe out in 1 minute.
This is your rate of breathing at rest.

2 Copy this table and put in your first reading:

Breathing rate at rest (breaths per minute)	Breathing rate after light exercise (breaths per minute)	Breathing rate after heavy exercise (breaths per minute)

3 Do step-ups for 1 minute (light exercise).
As soon as you have finished, sit down and count your breathing rate.
Put your reading in the table.

4 Now do step-ups as quickly as you can for 3 minutes (heavy exercise).
As soon as you have finished, sit down and count your breathing rate.
Put your reading in the table.

a What happened to your breathing rate in the investigation?
b Why do you think this has happened?
c What things do you think affect your rate of breathing?
d Did you notice any other changes to your body during the exercise?
   i) What about changes to your skin?
   ii) Did you feel hot?

Why do we have to breathe?

All the cells in your body need energy to stay alive.
Can you remember where you get your energy?

You get energy out of your food in respiration:

SUGAR + OXYGEN → CARBON DIOXIDE + WATER + ENERGY

Oxygen is needed for respiration to happen.
When sugar is burnt in oxygen, it gives out energy for your cells to use.
You get oxygen into your body by breathing it in.

▶ Use this information to explain why your breathing rate went up when you did more exercise.

How do you get oxygen into your body?

When you breathe in, air goes down your wind-pipe to your lungs.
Each lung is about the size of a rugby ball.

e Where do you think your lungs are found?

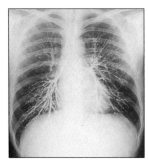

A chest X-ray

Look at the 2 diagrams.
The one at the top shows what the inside of your chest looks like.
The one at the bottom shows a chest model.

f Which part of the human chest is shown by the i) balloons?
ii) bell-jar? iii) rubber sheet? iv) glass tube?

g When you breathe in, do your lungs get bigger (inflate) or
smaller (deflate)?

h What happens to your lungs when you breathe out?

▶ Get the chest model.
Pull the rubber sheet down and then push it up.
Do this a few more times and watch what happens to the balloons.

i When you breathe in, does your **diaphragm** move up or down?

j What happens to it when you breathe out?

▶ Measure the size of your chest with a tape.
Now take in a deep breath.

k What happens to the size of your chest when you breathe in?

l What happens when you breathe out?

▶ Put your hands on your chest.
Breathe in and out deeply and slowly.

m Which way do your ribs move when you breathe in and out?

Muscles raise and lower your ribs and raise and lower your diaphragm.

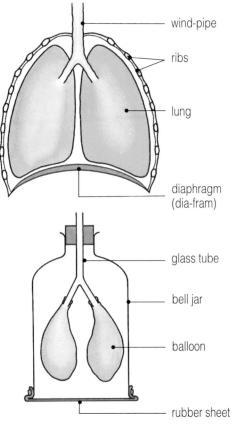

wind-pipe

ribs

lung

diaphragm
(dia-fram)

glass tube

bell jar

balloon

rubber sheet

---

1 Copy and complete the table using the information on this page.

	Breathing in	Breathing out
What do the ribs do? What does the diaphragm do? What happens to the space inside your chest? What happens to your lungs?		

Things to do

2 Try making your own 'model lungs'.
You could use an old plastic lemonade
bottle, balloons, a rubber-band and a plastic
drinking straw.
Make a hole in the top for the straw and
make it air-tight with plasticine. Cut away
the bottom of the bottle and stretch a
balloon over it for a diaphragm.

3 At the top of high mountains there is far
less oxygen in the air.
a) How do mountaineers manage to
breathe?
b) Why do you think many athletes train at
high altitude?

A change of air

"You breathe in oxygen and breathe out carbon dioxide" said Robert.
Do you think that he is right?

▶ Try breathing out through a straw into a test-tube of lime water.
What change did you see?
This is a test for carbon dioxide.

▶ Look at the pie-charts:

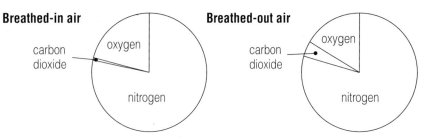

Breathed-in air

carbon dioxide — oxygen

nitrogen

Breathed-out air

carbon dioxide — oxygen

nitrogen

⚠
eye protection

a Which gas makes up most of the air you breathe in?
b Which gas do you breathe in more of?
c Which gas do you breathe out more of?

▶ See what happens to a candle when it is left to burn in
i) fresh air and ii) breathed-out air (your teacher will show you
how to collect this).
(Hint: a stop-watch might be helpful.)

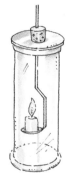

d What did you observe?
e Try to explain what you saw.

In and out ...

Set up the apparatus as shown in the diagram:

Breathe gently in and out of the mouth-piece several times.

f When you breathe in, does the air come in through A or through B?
g When you breathe out, does the air go out through A or through B?
h In which tube did the lime water turn cloudy first?

Write down your conclusions for your experiment.

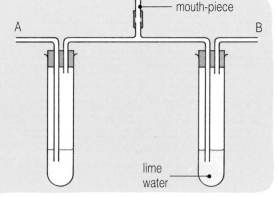

mouth-piece

A   B

lime
water

Gail said "You must have the same volume of lime water in each
tube".
Can you explain why she is right?

Look back at what Robert said at the top of the page.
Can you write out a better sentence?

Looking into your lungs

The diagram shows you how air gets to your lungs through your wind-pipe and then through the air passages.
The air passages end in tiny bags called **air sacs**. These have very thin walls. They are surrounded by lots of tiny **blood vessels**.

i How do you think oxygen gets from your lungs to all the cells of your body?

j How do you think carbon dioxide gets from the cells of your body to your lungs?

k Which gas do you think passes from the air sacs into the blood vessels?

l Which gas do you think passes in the opposite direction?

m These gases are swopped very quickly.
Write down 2 things that help this to happen (hint: read the sentences above again).

A lot of hot air

▶ Try breathing out onto a cold glass surface, like a beaker of cold water or a window.
What do you see?
Now put a strip of blue **cobalt chloride paper** onto the glass.
What happens?
Write down your conclusions.

Can you think of any other differences between the air you breathe in and the air you breathe out?
For one thing, the air you breathe out is *cleaner*.
Your air passages are lined with a slimy liquid called **mucus**. This traps dust and germs.
Then millions of microscopic **hairs** carry the mucus up to your nose and throat.

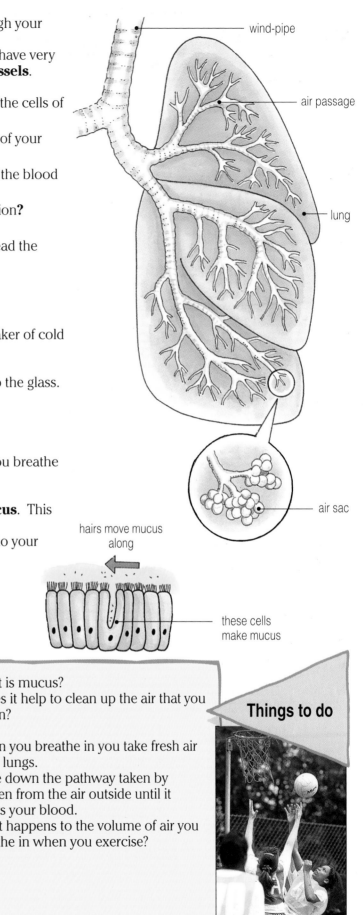

wind-pipe

air passage

lung

air sac

hairs move mucus along

these cells make mucus

---

1 Copy and complete:
We breathe in air containing nitrogen, ... and some carbon dioxide. The air that we breathe . . . . contains the same amount of . . . . , less . . . . and . . . . carbon dioxide. The air we breathe out also contains more . . . . vapour and is at a . . . . temperature.

2 Do you know about **artificial respiration**?
It's a way of starting up someone's breathing again.
You can learn about it at a first-aid class.
Your teacher can give you a Help Sheet explaining how it works.

3 What is mucus?
How does it help to clean up the air that you breathe in?

4 When you breathe in you take fresh air into your lungs.
a) Write down the pathway taken by oxygen from the air outside until it enters your blood.
b) What happens to the volume of air you breathe in when you exercise?

Things to do

Dying for a smoke?

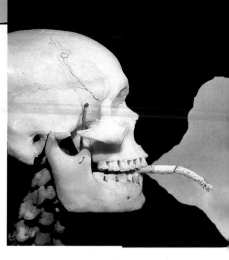

Do you know anyone who smokes?
There are fewer smokers around these days.
Lots of people used to smoke but nowadays they are finding it less attractive.

▶ Write down your ideas about why this is.

Did you know that cigarette smoke is made up of lots of chemicals and many of these are poisonous?
If you smoke, these chemicals go into your body through your mouth and along your air passages.

Fancy this lot?

NICOTINE R.I.P.
An addictive drug. It goes into your blood in the lungs. It causes your blood pressure to rise and your heart to beat faster.

TAR
A brown, sticky substance that collects in your lungs if you breathe in tobacco smoke. It is known to contain substances that cause cancer.

CARBON MONOXIDE R.I.P.
A poisonous gas. This prevents your blood from carrying as much oxygen as it should and so you get out of breath easily.

The smoking machine

First set up the apparatus without the cigarette.

Turn on the suction pump.

After 5 minutes, record:

- the temperature
- the colour of the glass wool
- the colour of the lime water.

Now repeat the experiment with a cigarette.

Record your observations.

What does this experiment tell you about the difference between breathing in fresh air and cigarette smoke?

thermometer
cigarette
to suction pump
rubber tubing
glass wool
lime water

Smoking changes people

teeth, fingers and nails turn yellow – that's the nicotine

smoker's cough – to get rid of the mucus

hair and clothes smell – that's the smoke

tongue turns yellow – you can't taste food properly

A mug's game

▶ Read what some people say about smoking.
Design a leaflet for primary school children to explain why they should not start smoking.

Smokers are 2 to 3 times more likely to die of a heart attack.

The money spent on cigarettes can't be spent on food, clothes etc.

Smoking increases the risk of serious diseases like bronchitis.

Ninety per cent of lung cancer occurs in smokers.

You can't keep the smoke to yourself. Everyone around you has to breathe it in.

Illnesses caused by smoking have to be treated. If people did not smoke, this would cost the country less money.

Smoking makes you breathless and less good at sport.

So why start?

Amy is 13. She smokes about 5 cigarettes a day.

▶ Read what Amy has to say about smoking:

"I had my first cigarette when I was 10.
My friend Sharon, she's 2 years older than me, offered me one.
I didn't like it much at first, but it felt exciting somehow.
My Mum would have killed me if she'd found out.
She's always trying to get my Dad to give up but he can't.
I suppose I spend about £3 or £4 a week on cigarettes.
I'll give it up when I'm older because it affects your health.
I'd definitely give up if I ever got pregnant."

In your groups, discuss the reasons why you think people start to smoke.

1 Copy and complete:
Cigarette smoke contains poisonous . . . .
One is a drug called . . . . This gets into your blood in the . . . . It causes your blood . . . . to rise and your heart to beat . . . . Tar contains chemicals that cause . . . . A poisonous gas called . . . . stops your blood from carrying as much . . . . as it should.

2 You are asked to talk to some junior school children about the dangers of smoking.
Plan out what you are going to tell them.

3 Write down what you think about the following:
a) Smoking should be banned in shops and offices.
b) Once you start smoking it's hard to stop.
c) Smoking costs us all a lot of money.

4 The graph shows how the risk of getting lung cancer changes after giving up smoking. Explain why you think smokers should give up the habit.

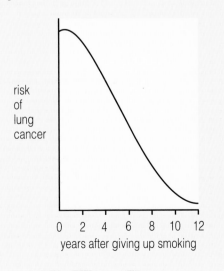

risk of lung cancer

years after giving up smoking

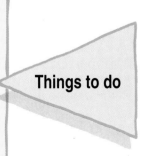

Things to do

Living liquid

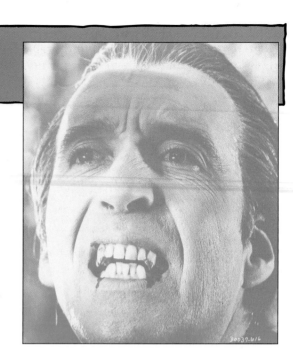

What does blood make you think of?
Horror movies, vampires, wars?

You probably have about 4 litres of blood in your body. That's a bucket-full.
It's flowing around your body all the time.
But what is it for?

▶ Write down your ideas about how your blood helps you.

You already know that the cells of your body need food and oxygen to give you energy in respiration.
Your cells make waste chemicals too.
Your kidneys get rid of most of these waste chemicals.

▶ Look at the diagram and answer the questions:

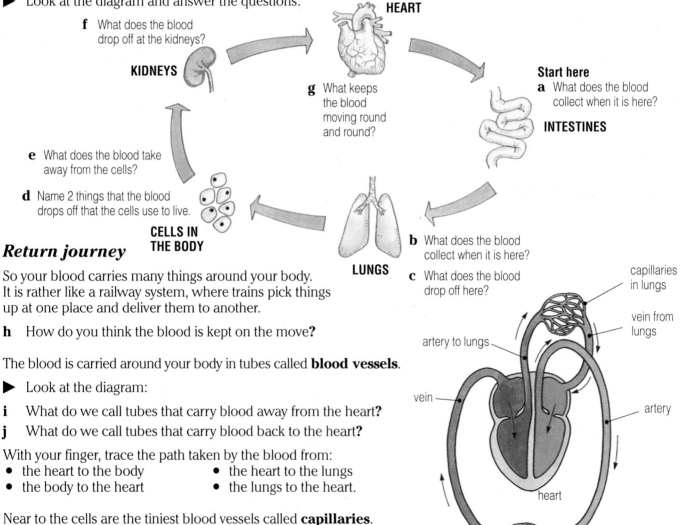

HEART

f What does the blood drop off at the kidneys?

KIDNEYS

g What keeps the blood moving round and round?

Start here
a What does the blood collect when it is here?

INTESTINES

e What does the blood take away from the cells?

d Name 2 things that the blood drops off that the cells use to live.

CELLS IN THE BODY

LUNGS

b What does the blood collect when it is here?

c What does the blood drop off here?

capillaries in lungs

vein from lungs

artery to lungs

vein

artery

heart

capillaries in body

Return journey

So your blood carries many things around your body.
It is rather like a railway system, where trains pick things up at one place and deliver them to another.

h How do you think the blood is kept on the move?

The blood is carried around your body in tubes called **blood vessels**.

▶ Look at the diagram:

i What do we call tubes that carry blood away from the heart?
j What do we call tubes that carry blood back to the heart?

With your finger, trace the path taken by the blood from:
● the heart to the body   ● the heart to the lungs
● the body to the heart   ● the lungs to the heart.

Near to the cells are the tiniest blood vessels called **capillaries**.
These connect your arteries to your veins.

k Why do you think that capillaries have very thin walls?

Around and around . . .

Do you remember how to take your pulse?
When you take your pulse you feel an artery.
Blood flows through your arteries in spurts.
This is the pulse that you feel.

Does your pulse rate and breathing rate go up and down together?

Plan an investigation to see how your pulse and breathing are affected by exercise.

- What exercise will you plan to do?
- What measurements will you take?
- How will you make it a fair test?
- How do you plan to show your results?
- Check your plan with your teacher before carrying it out.

Supply lines

▶ Lift one arm above your head and let the other arm hang at your side. Keep them there for a minute or two.
Now bring them in front of you and look at the differences between the veins on the back of each hand.
What differences can you see?

▶ Look at the photograph of a section of an artery and a vein.

l   What differences can you see?

m   Can you find out any other differences between arteries and veins?

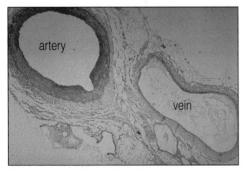

Section of an artery and a vein

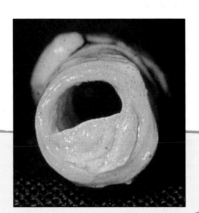

Things to do

1   Copy and complete:
Blood is pumped around my body by my
. . . . Blood travels away from my heart in
. . . . and back to my heart in . . . . The tiniest
blood vessels are called . . . . and these have
very . . . . walls so things can pass in and out.
When I feel my pulse I am touching an . . . .

2   Make a table of the differences between arteries and veins.

3   Sometimes our arteries can get 'furred up'. This is because a fatty substance sticks to the inside of the artery and makes it narrower. How do you think this would affect the flow of blood in the artery?

Heartbeat

18e

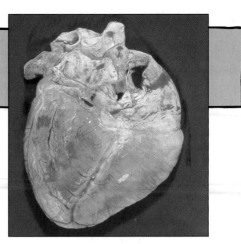

A human heart

What do you think is the strongest muscle in your body**?**
Not many people think that it is their heart.
Just think of the job your heart does.
It beats about 70 times a minute, for 60 minutes per hour and
24 hours a day, to keep you alive.

a   Use a calculator to work out how many times your heart beats:
i) per hour   ii) per day   iii) in a year.

The double pump

Where do you think your heart is**?**
Put your hand on the place where you think it is.
What can you feel**?**

b   How do you think your heart is protected**?**

c   How many spaces are there inside your heart**?**

▶   Look at the diagram. It is drawn as though you are facing
someone.

Your heart is really 2 separate pumps side-by-side.
When your heart beats, the muscle squeezes the blood out.

d   Where does the right-side of your heart pump the blood to**?**

e   Where does the left-side of your heart pump the blood to**?**

f   Which side of the heart will have blood containing most oxygen**?**

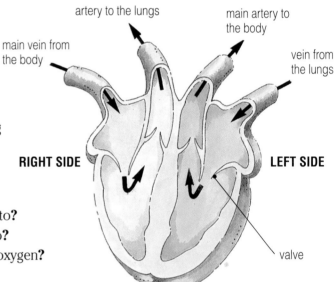

artery to the lungs

main artery to
the body

main vein from
the body

vein from
the lungs

RIGHT SIDE

LEFT SIDE

valve

Heart listening

The 2 pumps both beat at the same time.
You can hear your partner's heartbeat by using a **stethoscope**.
Try making a home-made version as shown in the diagram.
What sounds can you hear**?**

You should hear 2 sounds.
Doctors call them lub-dub sounds.
The 2 noises are caused by the **valves** in your heart closing.
Try listening again . . . lub-dub . . . lub-dub . . . lub-dub . . .

g   Why do you think your heart valves close**?** (Hint: look at the
top diagram.)

Find out if your heartbeat (the number of beats per minute) is the
same as your pulse rate.

h   Does it change in the same way as your pulse rate when you
exercise**?**

A home-made stethoscope

The big killer

Heart disease is one of the biggest killers in Britain.
Fatty substances can 'fur up' the arteries leading to the heart muscle.

i   What would 'furring up' do to the flow of blood to the heart muscle?

j   What could happen to the supply of oxygen to the heart muscle?

If the heart muscle does not get enough oxygen it can cause chest pains.
This is called **angina**.
It is a warning that the person is more likely to have a heart attack.

Sometimes a clot can form inside a **coronary artery**.

k   How do you think this could cause a **coronary heart attack?**

▶   Look at the cartoons:
In your groups discuss the things that you think can increase the risk of heart attack.
How do you think each of these risks could be reduced?
Make a list of your ideas.

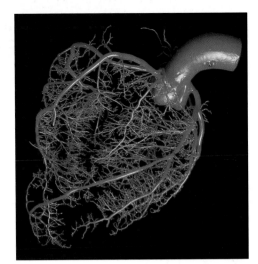

A cast of the coronary arteries

1   Copy and complete:
The heart is made out of . . . . The blood on the left-side contains . . . . oxygen than the blood on the . . . . -side. This is because the blood has just come back from the . . . . The left-side of the heart pumps blood all around the . . . . The heart has . . . . to stop the blood from flowing backwards.

2   Look at the heart diagram opposite. List what happens to the blood as it passes from the main vein to the heart and eventually to the main artery.

3   What sort of person do you think is most likely to suffer a heart attack? How old will they be? What is their weight like? What sort of habits might they have? Draw a cartoon of the person and label it.

4   Design a leaflet or make a poster to let people know about the risks of heart disease.

Things to do

Liquid protection

Do you know what a **blood transfusion** is?
It once saved Richard's life.
He was involved in a motorway accident.
He was losing a lot of blood.
Fortunately for him, the ambulance team
were quick to arrive. So they were able to
give him extra blood.
But not all our blood is the same.
Richard carried a card that showed which
type his blood was.

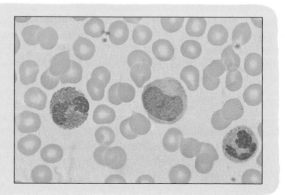

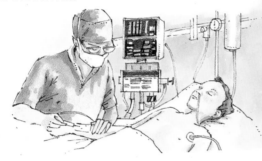

a There are 4 main blood types. Do you know what they are called?

Each blood type contains slightly different chemicals.

b Why do you think that you can only have a transfusion of blood
of the same type?

Richard was very grateful to the blood donors who gave their blood.

c What do you have to do to be a blood donor?

Blood bank

Blood taken from a donor is treated so that it does not clot.
It may settle out into 2 parts. A pale yellow liquid called **plasma**
and a deep red layer of **blood cells**.

d Is the blood made up mainly of plasma or blood cells?

Plasma is mainly water containing dissolved chemicals.

Your teacher will give you a prepared slide of blood cells. Look at it
under your microscope. You will need to focus carefully at high
power.

e Which do you think are bigger: red cells or white cells?

f Are there more red cells or white cells?

Draw a diagram of each different blood cell that you can see.

g Can you see any other differences between red and white cells?

h Why do you think the white cells look purple?

The oxygen carriers

The red cells carry oxygen.
They are red because they contain **haemoglobin**.
This is a chemical that can collect and carry oxygen.
Haemoglobin lets go of oxygen when it comes to a part of the body
that needs it.

i   Where do you think haemoglobin collects oxygen from**?**

▶ Copy this diagram. Add as many labels and notes to it as you
   can to explain how oxygen is carried round your body.

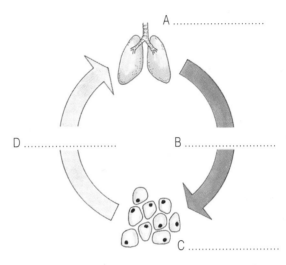

A ....

D ....   B ....

C ....

The protection gang

White cells protect your body from germs.

j   In what ways could germs get into your body**?**

One sort of white cell *eats* any germs that it finds.
Another sort makes chemicals (called **antibodies**) that can kill
germs.

k   Can you see 2 types of white cell in the photograph at the
    bottom of page 74**?**
    In what ways do they look different**?**

l   What happens just after you cut yourself**?**

Eventually a scab will form.
But first the bleeding has to be stopped.
The cut is sealed by lots of tiny bits of cells called **platelets**.

m   Why is it important that the cut is sealed quickly**?**

1   Copy and complete:
The liquid part of the blood is called the . . . .
The red cells contain a substance called . . . .
This substance helps the . . . . cells to carry
. . . . The white cells . . . . the body from
germs. One kind of white cell . . . . germs.
Another type makes . . . . that kill germs.

2   Make a table of the differences between
red and white blood cells.

3   People who live at high altitude have far
more red cells than you. Why do you think
this is?

4   These days in science lessons you are
not allowed to take blood samples. Why do
you think this is?

Things to do

Questions

1 Gemma and Beth measured their breathing rate (in breaths per minute) before they ran a race. Then they measured their breathing rates again, every minute, until their rates returned to normal. They recorded their results in a table:

	Before exercise	Minutes after exercise						
		1	2	3	4	5	6	7
Gemma	16	45	38	31	24	20	17	16
Beth	13	35	32	28	22	18	13	13

   a) Plot 2 line-graphs on the same sheet. Use the vertical axis for breathing rate and the horizontal axis for time.
   b) Who took the longer time to recover from the exercise?
   c) Who do you think was the fitter of the two girls? Give your reasons.

2 Where in your body are the following found:
   a) diaphragm?   b) air sacs?   c) valves?   d) capillaries?
What job does each one do?

3 Do you think the following statements are true or false?
   a) Most of the air you breathe is not used by your body.
   b) Smoking does not increase the risk of heart disease.
   c) You can get lung cancer by breathing in other people's cigarette smoke.
   d) Exercise increases the risk of heart disease.

4 Blood has many different jobs.
Which part of your blood:
   a) carries oxygen?   b) carries dissolved food?
   c) fights germs?   d) helps your blood to clot?

5 Coronary heart disease can be caused by eating too much saturated fat in foods like meat, butter and cream. You should choose unsaturated fats, in foods like fish and vegetable oils.
Look at food packets and make a list of foods for each of these 2 types of fat.

6 a) Find out where your red blood cells are made.
   b) Find out how lack of iron causes **anaemia** and how this affects the body.
   c) Why do you think that women take more iron tablets than men?

7 Look at this blood transfusion certificate:
Which blood group does this person belong to?
Make a leaflet or a poster that will persuade people to give blood.

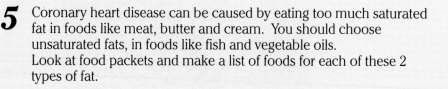

Sight and Sound

19

Seeing and hearing are important to us.
We need them to communicate with our friends.
Have you ever thought what it would be like to be blind
and deaf?

In this topic you can learn more about how we use light
and sound waves.

Bending light

▶ Look at this picture:

A ray of light from the lamp is **reflected** from the mirror.

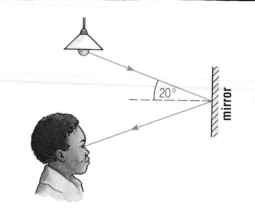

a Which is the incident ray?

b Which is the reflected ray?

c If the angle of incidence is 20°, how big is the angle of reflection?

d Why does the boy see the lamp?

Reflection from a mirror is one way of changing the direction of a ray. Here is another way, using **refraction**.

Investigating refraction

Your teacher will give you a Help Sheet and a semi-circular block of glass (or perspex).

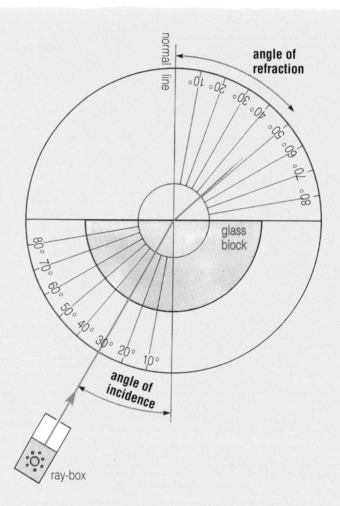

1 Place your block of glass in position on the Help Sheet, as shown here:

2 Use a ray-box to send a thin beam of light at an angle of incidence of 30°, as shown.

3 Look carefully at the ray where it comes out of the block.
Can you see the ray does not go straight on? It changes direction at the surface of the block. We say the ray is **refracted**. This is **refraction**.

4 Mark on the paper the path of the ray coming out of the block. Label it ray ①.

5 Now increase the angle of incidence to 40°. What happens?
Mark the new path of the refracted ray, and label it ②.

You can see that when light comes out of a glass block, it is bent (refracted) *away from* the normal line.

6 Now increase the angle of incidence to 50°. What happens now?
Mark the new path of the ray, and label it ③.

When the angle of incidence is large, you can see that the light is reflected *inside* the glass. We call this **total internal reflection**. This can be very useful, as you'll see on the next page.

Using refraction

A **lens** is a shaped piece of glass. There are two kinds:

A *convex* lens is fat in the middle.
A *concave* lens is thin in the middle.

When light goes through a lens it is refracted.

A convex lens brings the rays of light closer together.
We say they are **converging**.

A concave lens makes the rays spread out. They are
diverging.

The rays always bend towards the thickest part of a lens.

e Where is there a lens in your body?

f Is it convex or concave?

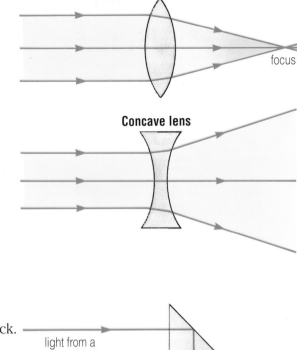

Convex lens

focus

Concave lens

Using total internal reflection

You discovered that light can be reflected *inside* your glass block.
This is used in the **cat's eye** reflector that you see on the roads.
The cat's eye is a triangular piece of glass, called a **prism**:

Light from a car head-lamp is reflected twice inside the prism, and
then shines back into the driver's eye.

light from a
car headlight

Cat's eye
reflector

Doctors can use total internal reflection to see into your stomach.
They use a long thin piece of glass called an **optical fibre**.
The light is reflected from side to side along the glass fibre:

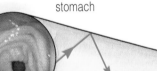

stomach

flexible optical fibre

1 Copy and complete:
a) When light goes in or out of glass, it . . . .
direction. The rays are . . . .
This is called . . . .
b) When light comes out of glass it is . . . .
away from the normal line. When it goes
into glass it is . . . . towards the normal.
c) In a convex lens the . . . . come closer
together. The rays are . . . .
d) In a . . . . lens the rays spread out. They
are . . . .
e) Total . . . . reflection is used in cat's eye
. . . . and in . . . . fibres.

2 Look at the diagram of a **con***vex* lens at
the top of this page.
Draw similar diagrams to show what you
think would happen to the rays if the lens
was: a) fatter,   b) thinner.

3 Make a list of all the things you can
think of that use a lens.

4 If you had an optical fibre several
metres long, how could you use it to send
messages to a friend in another room?
What other uses can you think of?

Things to do

A world of colour

a   Why do you think road signs are often coloured red?

b   Imagine a world without colour. Describe what it would be like to live in it.

c   Write down what you think are the colours in a rainbow.

The colours in a rainbow form a **spectrum**.

Making a spectrum

Shine some white light from a ray-box through a prism, and on to a screen:

Turn the prism until you see a spectrum on the screen.

d   How many different colours can you count?

e   Which colour has been bent the least?

f   Which colour has been refracted the most?

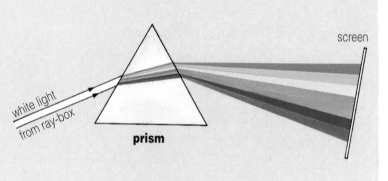

This experiment shows that white light is really a mixture of several colours. The colours are split up by the prism.

We say that the white light has been *dispersed* by the prism to form a visible spectrum. This is **dispersion**.

The colours, in order, are: **R**ed, **O**range, **Y**ellow, **G**reen, **B**lue, **I**ndigo, **V**iolet.
You can remember them as a boy's name: **ROY G. BIV**.

Light waves

If you throw a stone in a pond, you can see the ripples or **waves** spreading out from it.
Light spreads out in the same way, in waves.

Each wave has its own **wavelength**.
Different colours have different wavelengths.

Red light has the longest wavelength. It is about $\frac{1}{1000}$ mm.

Violet light has the shortest wavelength. There are about 2000 wavelengths of violet light in 1 mm.

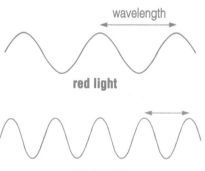

Seeing coloured objects

When light shines on a coloured object, some of the light is taken in or **absorbed**.
The rest of the light is reflected.
You see the colour of this reflected light. For example:

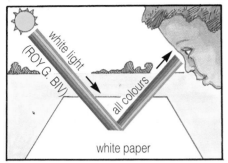

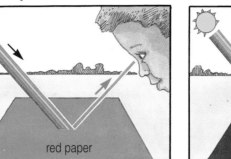

White things reflect all of the colours of light. That is, all of ROY G. BIV.

Red things reflect red light and absorb the other colours. We see the red light.

Black things do not reflect any light. All the light is absorbed.

g What colour light is reflected by a blue T-shirt?

h Explain what is happening when you look at this red ink

Looking at objects in different colours of light

Plan an investigation that will give you the data for this table:

- To make the coloured light you can use a coloured **filter** on a ray-box (or torch).
  A red filter lets only red light through it.

- Show your plan to your teacher, and then do it.

- What pattern do you find?

	Colour of objects in coloured lights			
Colour of object in daylight	Colour of light shining on it			
	white	red	green	blue
white	white			
red				
green				
blue				

Colourful cars

Plan an investigation to see **which is the safest colour for a car**.
That is, which colour can be most easily seen in a) daylight,
b) street lights, and
c) headlights.

Show your plan to your teacher, and if you have time, do it.

1 Copy and complete:
a) .... light is a mixture of 7 colours.
b) The 7 colours of the spectrum are:
c) .... light has the longest wavelength.
d) A red T-shirt reflects .... light and ....
all the other colours.

2 What colour would a blue book look:
a) in white light?   c) in red light?
b) in blue light?   d) through a red filter?

3 Imagine you are in a rock band, and your stage lights usually flash red or blue.
Design some clothes which will look good in red, in blue and in white light.
Draw colour pictures of what they will look like in: a) white, b) red, c) blue light.

4 What is camouflage? Draw some camouflage suitable for a bird-watcher
a) in the desert,   b) in the jungle.

Things to do

81

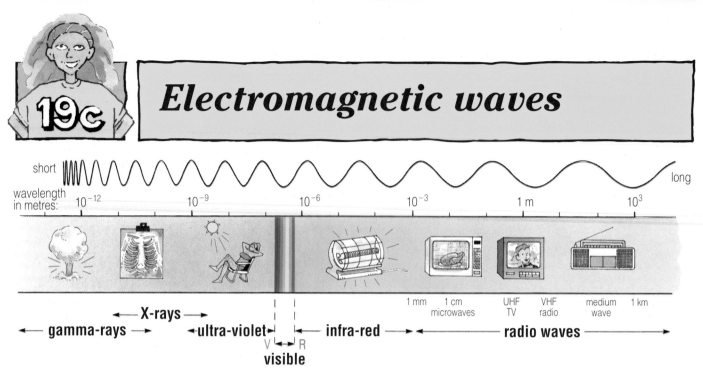

short					long
wavelength in metres:	10^{-12}	10^{-9}	10^{-6}	10^{-3}	1 m	10^3

1 mm 1 cm UHF VHF medium 1 km
microwaves TV radio wave

←— X-rays —→
←— gamma-rays —→ ←ultra-violet→ ←— infra-red —→ ←————— radio waves —————→
V ←→ R
visible

a Write down the 7 colours of the visible spectrum.

The spectrum that you can see is just a small part of a much bigger spectrum, called the **electro-magnetic spectrum**. This is made of several different types of *radiation*.

The full electro-magnetic spectrum is shown in the diagram above. Study it carefully.

b Write down the 6 main types of radiation, starting with gamma-rays.

c Which type of radiation has the longest wavelength**?**

d Which type of radiation has the shortest wavelength**?**

Now use the data given on these two pages to answer questions **e** to **n**.

Electromagnetic waves:

1 All can travel through empty space (a vacuum).

2 All travel at the same speed as light: 300 000 kilometres in every second.

3 All can be reflected and refracted.

4 They all transfer energy from one place to another.

5 The shorter the wavelength, the more dangerous they are.

Gamma-rays (γ-rays) have the shortest wavelength. They are very penetrating and can pass right through metals.

They are given out by radio-active substances. They are very dangerous to humans unless used carefully.
Gamma-rays are used to kill bacteria and sterilize hospital equipment.
With careful control they can be used to kill cancer cells. This is called radiotherapy:

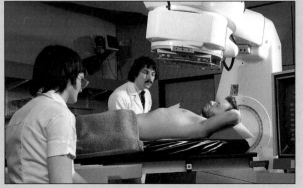

X-rays are very like gamma-rays. They have a short wavelength.

X-rays are high-energy waves and are very penetrating. They can pass through skin easily, but not so easily through bones. Doctors and dentists use them to check bones and teeth:

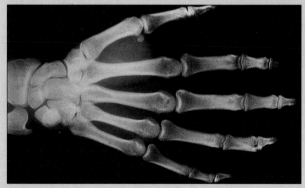

X-rays can be dangerous, because they can damage cells deep inside your body.
Pregnant women should avoid X-rays because they can damage the baby.

Ultra-violet rays (UV rays) come from the Sun and from sun-lamps. These are the waves that give you a sun-tan.

Ultra-violet rays damage your skin. They can cause skin cancer. This can be prevented by using sun-cream to block out the harmful rays.

UV can also be used to kill bacteria.

Infra-red rays (IR rays) are waves that are longer than red light. You cannot see IR rays but you can feel them on your skin as warmth.
Any warm or hot object gives out infra-red rays – including you!

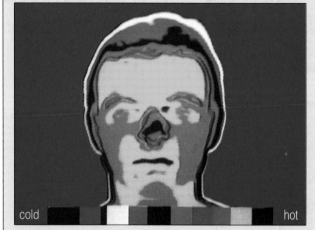

cold   hot

Fire-fighters use infra-red detectors to look for people in smoke-filled rooms, and to find people buried alive after an earthquake.

e Which types of radiation cannot be seen by our eyes**?**

f Which rays are very like X-rays**?**

g Name a type of electromagnetic wave which:
   i) Can cause a sun-tan.
   ii) Can pass through metals.
   iii) Is emitted by warm objects.
   iv) Can be seen by your eye.

h Which waves are often used for communication**?**

i Are there radio waves in this room at the moment**?** How could you find out**?**

j How far do radio waves travel in 1 second**?**

k Which rays can be harmful to life**?**

l Write a newspaper article on sun-bathing. Discuss why people do it, and its dangers.

m Draw a poster to warn of the dangers of X-rays.

n Imagine you are an alien, just landed on Earth. Your eyes can only see with infra-red radiation. Describe what you might see as you leave your space-ship and walk through a town.

Radio waves have a longer wavelength.
There are several kinds of radio waves:

Microwaves are used in microwave ovens. The energy of the microwaves heats up the food. Microwaves are also used for radar and for communicating with satellites.

UHF waves (**U**ltra **H**igh **F**requency) are used to transmit TV programmes to your home.

VHF waves (**V**ery **H**igh **F**requency) are used for local radio programmes, and by the police.

Medium wave and **long wave** radio are used to transmit over long distances.

---

1 Copy and complete:
a) The full electromagnetic spectrum, in order, is gamma-rays, ... , ... , ... , ... , ...
b) . . . . waves have the longest wavelength.

2 Write down 5 things that electromagnetic waves have in common.

3 Cut out pictures from magazines to make a collage of the uses of different types of radiation.

4 Rattlesnakes have an extra pair of 'eyes' which see infra-red rays. Explain how this helps the snake to catch food. Why would the snake find it easier at night?

5 What do microwaves do to food? What materials can microwaves:
a) pass through?   b) not pass through?

6 Garry says that all 6 types of radiation are used somewhere in a hospital. Do you agree? Explain your answer.

Things to do

Sound moves

▶ Look at the picture:

a Jan's throat is vibrating. Explain, step by step, how Mei hears the sound waves.

▶ Rabbits warn each other of danger by thumping the ground.

b Do you think sound can travel through a solid?

c How do whales 'talk' to each other?

d Can sound travel through water?

Sound travels

Use a 'slinky' spring to show how sound moves.

Push one end of the spring to compress it:

Then pull it back:

Then push it again, to compress it:

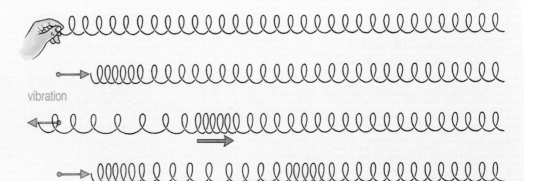

This kind of wave is called a longitudinal wave.

You are **vibrating** the end of the spring.
A **wave** of energy travels down the spring.

You can see that some parts of the spring are pushed together, and other parts are pulled apart.

When you speak, sound energy travels from your mouth in the same way.
Instead of a spring, there are air particles called **molecules**.
When you speak, the molecules are pushed together and pulled apart, so that they vibrate like the spring.
The sound energy travels away from your mouth, like the wave on the spring. Sound travels 340 metres in every second.

Can sound travel through a vacuum?

e What is a vacuum?

Your teacher will show you what happens to sound when the air is removed from a bell-jar:

f First, **predict** what you expect to happen.

g What happens as the air is pumped out?

h What happens as the air is let back in?

i Explain this experiment.

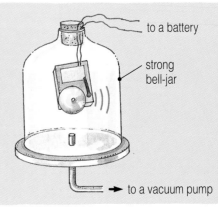

to a battery

strong bell-jar

to a vacuum pump

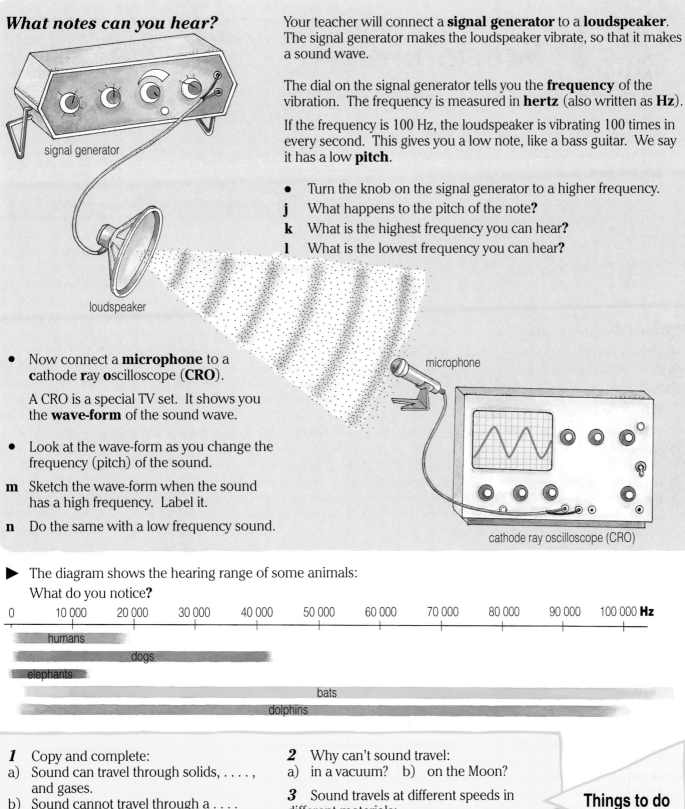

What notes can you hear?

Your teacher will connect a **signal generator** to a **loudspeaker**. The signal generator makes the loudspeaker vibrate, so that it makes a sound wave.

The dial on the signal generator tells you the **frequency** of the vibration. The frequency is measured in **hertz** (also written as **Hz**).

If the frequency is 100 Hz, the loudspeaker is vibrating 100 times in every second. This gives you a low note, like a bass guitar. We say it has a low **pitch**.

signal generator

- Turn the knob on the signal generator to a higher frequency.
- **j** What happens to the pitch of the note?
- **k** What is the highest frequency you can hear?
- **l** What is the lowest frequency you can hear?

loudspeaker

- Now connect a **microphone** to a **c**athode **r**ay **o**scilloscope (**CRO**).

  A CRO is a special TV set. It shows you the **wave-form** of the sound wave.

- Look at the wave-form as you change the frequency (pitch) of the sound.

- **m** Sketch the wave-form when the sound has a high frequency. Label it.

- **n** Do the same with a low frequency sound.

microphone

cathode ray oscilloscope (CRO)

▶ The diagram shows the hearing range of some animals:
What do you notice?

| 0 | 10 000 | 20 000 | 30 000 | 40 000 | 50 000 | 60 000 | 70 000 | 80 000 | 90 000 | 100 000 **Hz** |

humans

dogs

elephants

bats

dolphins

Things to do

1 Copy and complete:
a) Sound can travel through solids, ... , and gases.
b) Sound cannot travel through a . . . .
c) In a sound wave, the tiny . . . . are pushed together and pulled farther apart.
d) The frequency of a vibration is measured in . . . . This is also written . . . .
e) A note with a high pitch has a high . . . .
f) The human range of hearing is from about 30 Hz up to about . . . . Hz.

2 Why can't sound travel:
a) in a vacuum?   b) on the Moon?

3 Sound travels at different speeds in different materials:

Material	air	water	wood	iron
Speed of sound (m/s)	340	1500	4000	5000

a) Plot a bar-chart of this information.
b) What pattern can you see?

Noise annoys

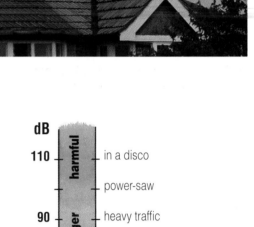

▶ What do you think is the difference between *music* and *noise?*

a Write down 3 words to describe music.
b Write down 3 words to describe noise.

Noise is any sound that we don't like. It is a kind of pollution.

c Give 3 examples of sounds that are noise to you.

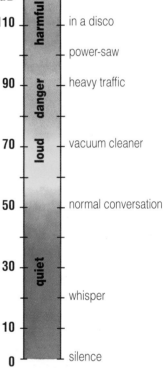

The loudness of any sound is measured in **decibels**. This is often written as **dB**.

The quietest sound anyone could hear is zero decibels (0 dB). A louder sound, with more decibels, has more *energy*.

The scale shows the loudness of different sounds:

d What is the loudness, in dB, of a normal conversation?

Try to estimate the loudness (in dB) of these sounds,
e birds singing,
f a food-mixer.

Loud sounds are dangerous. They can make you permanently deaf.

g If you were using a power-saw, what should you wear?

h Why is it often dangerous in a disco?

i Why does the noise in a sports hall sound loud?
j *Why* would it change if there was a carpet and curtains?

Scale (dB):
- 110 — harmful — in a disco
- power-saw
- 90 — danger — heavy traffic
- 70 — loud — vacuum cleaner
- 50 — normal conversation
- 30 — quiet
- whisper
- 10
- 0 — silence

Investigating sound-levels

Use a sound-level meter to measure the loudness of some sounds:

Show your results on a scale like the one above.

Using a sound-level meter   Wearing ear-defenders at work

Noisy neighbours

Kelly's neighbours are very noisy. The noise comes through the walls while Kelly is doing her homework. She wants to make the walls of her bedroom sound-proof.

Plan an investigation to find out **which material is best for making sounds quieter**.

materials

- Your teacher will give you several materials.
  For example: paper, kitchen foil, foam rubber, cloth, polythene bag, plasticine, sellotape, cotton wool, etc.

sources

- You will need something to make the sound.
  For example: a clock or a buzzer, or a tin-can containing stones, or a radio, or your voice.

detectors

- You will need something to detect how much sound gets through the material.
  For example: your ear, or a sound-level meter, or a microphone and CRO.

- Plan what you are going to do. How will you make it a **fair test?**

- How will you record your results**?**

- Show your plan to your teacher, and then do it.

- Write a report for Kelly, explaining which material is best for sound-proofing her room.

- Which sorts of materials are best**?** Is there any pattern in your results**?**

- How could you improve your test**?**

Imagine you live on a very busy road, with a lot of traffic noise.
In your group, discuss these questions:

k How could you cut down the traffic noise in your garden**?**

l How could you cut down traffic noise inside your house**?**

m What do you think should be done to reduce traffic noise in towns**?**

---

1 Copy and complete:
a) The loudness of a sound is measured in . . . . (often written as . . . .).
b) A louder sound has more . . . .

2 Design a poster to encourage teenagers to protect their ears.

3 Imagine you are working in a noisy office. What suggestions could you make to improve it?

4 You are looking out of the window, in a dark room, drinking a cup of coffee, and with the TV on. A deaf person comes into the room. What things would you do before speaking to her?

5 Dave says, "I can work better on my homework if I have some music playing." Wayne says, "I don't agree – the music will spoil your concentration."
Plan an investigation to see who is right.

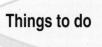

Things to do

Questions

1 A bike reflector is made of triangular pieces of red plastic. It works like a cat's eye reflector.
   a) Draw a diagram to show exactly how it reflects the light from the headlights of a car.
   b) Why is the reflected light coloured red?

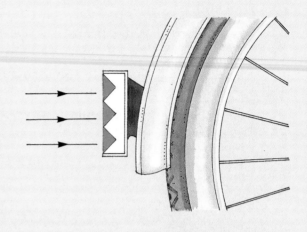

2 Some pupils were hypothesizing about the effect of colour.

   Andy said, "I think fewer people choose to eat green sweets than any other colour."
   Becky said, "I think birds prefer to eat white bread rather than brown bread."

   Choose one of these hypotheses, and plan an investigation to test it. Take care to make sure it is a fair test.

3 Make a colour survey of:
   either the clothes in your wardrobe, *or* the furnishings in your house, *or* the cars on your street.
   What is the best way to display your data?

4 Draw a poster warning people of the dangers of too much sun-bathing.

5 Carry out a sound survey among your family and friends. Make a list of everyone's favourite and least favourite sounds.

6 a) Copy out the table. Tick a column to show if the frequency is a high or a low pitch.
   b) Older people cannot hear very high notes. Why do you think your range of hearing will get smaller as you get older?
   c) What is ultra-sound? How do bats use it?

Frequency	Pitch	
	high	low
10 000 Hz		
50 Hz		
50 kHz		
20 kHz		

7 Alan, Bev and Claire have read that people find high frequency sounds more annoying than low frequency sounds. They each have a hypothesis:

   Alan says, "I think it's because high frequency sounds make it harder to hear someone talking."

   Bev says, "I think it's because our ears are more sensitive to high-pitched sounds."

   Claire says, "I think it's because high frequency sounds can penetrate through walls more easily."

   a) Do you agree with any of these hypotheses?
   b) Choose one, and plan an investigation to test it.

Water

20

Water keeps us alive. As human beings we need to drink it. We also need it to grow our food.

Next time you turn on a tap for a drink, think about where the water comes from.

In this topic you can find out why we should be pleased when it rains!

In this topic:
- **20a** *Water, water everywhere*
- **20b** *The water cycle*
- **20c** *What's in a cloud?*
- **20d** *Why does it rain?*

20a
Water, water everywhere

Water is a **compound**. It is made from 2 elements.
Can you remember which elements combine to make water?
(Hint: look at page 27.)

Water has different states.

ice   liquid water   steam

▶ Copy out these states. Add the following words in the right places:

evaporates   freezes   condenses   melts

▶ Your teacher will show you what *1 litre* of water looks like.
How much water does your family use each day? Estimate the
number of litres.
Think about the water your family uses for:

- flushing toilets
- washing people (baths, showers, . . .)
- washing clothes
- washing dishes
- cooking
- drinking.

Are there other uses?

Now do a rough calculation to estimate how much water is
used by all the families of people in your class.

▶ In many other countries water is in short supply.
Why do you think this is?
Many people have to carry the water they need for miles. How
do you think this affects the amount of water they use?
(Think about carrying a bucket of water around school for a few
hours . . .)

Testing water

What do you think pure water is? What is the difference between impure water and pure water?

Do the following tests on your samples of pure and impure water. Record your results in a table like this:
For each sample of water,

1 Measure its boiling point.
2 Measure its freezing point.
3 See if it conducts electricity.
4 Put a drop of the sample onto a piece of blue cobalt chloride paper. Note what you see.
5 Put a drop of the sample onto a piece of universal indicator paper. What is the pH number?

⚠️
eye
protection

Test	Pure water	Impure water
1. Boiling point		
2. Freezing point		
3.		

Which of the tests gives the same result for pure water and impure water?

I think it's best to wash my hair in rainwater. It's very pure.

I think water from streams is purer.

Plan an investigation to find out whether rainwater or stream water is purer. Check your plan with your teacher who may then let you carry it out.
What might cause any impurities in a) rainwater? b) stream water?

Things to do

1 Copy and complete the following sentences:
a) The melting point of pure water is . . . . °C.
b) The boiling point of pure water is . . . . °C.
c) When . . . . melts it forms liquid water.
d) When liquid water . . . . it forms steam.
e) Pure water can. . . . conduct electricity.
f) The pH of pure water is . . . .

2 Draw diagrams to show the arrangement of particles in the 3 states of water.

3 Design a poster to encourage people to use less water.
Why might it be important to save water?

4 Imagine that you have the job of selling water. Think about its properties and possible uses. Make an advertising leaflet to help sell the water.

5 FIZZO is a popular new blackcurrant flavoured drink. One of its ingredients is water.
a) Describe an experiment to get pure water from the drink.
b) Describe an experiment to find out whether the purple colouring is a pure, single substance or a mixture.

FIZZO
BLACKCURRANT DRINK

The water cycle

What states of water can you see in each of the pictures?

▶ Water is one of the most important substances on the Earth.

Do you agree with Jack?
Write about 4 or 5 lines to explain why.

We use billions of litres of water every day.
Have you ever wondered why we don't run out of it?
The answer is that it is **recycled**.

The water cycle is a very important process in nature.

In your group make a poster to show the water cycle.
This picture could be the *start* of your poster.

The labels give information about some parts of the water cycle.
Add them to your picture in the right places.
Where is the rain likely to fall?
Draw it on your picture.
Make sure you put arrows on your picture to show which way the water is moving.

a Energy from the Sun is very important to the water cycle. Which part of the cycle does this affect?

b Why does water sometimes fall as *snow* rather than rain?

c What causes acid rain?

Labels
- water evaporates from the sea
- water as a gas (vapour) moves upwards
- as it gets colder, water condenses into droplets to form clouds
- wind moves clouds
- droplets get heavier, and water falls as rain, hail or snow
- the water returns to the sea through lakes and rivers

Before water goes back to the sea it is piped into our homes from lakes, rivers or from underground.
You couldn't drink water which came straight from rivers.

Why not?

All water must be cleaned before we can use it.
This happens at a water treatment plant (waterworks).
Very large amounts of water must be cleaned cheaply.
The process used is called **filtration**.

Would you like to drink this?

Cleaning water

Your teacher will give you some muddy water.
Your task is to get the cleanest water you can from the muddy water.
You can use any of the equipment your teacher will give you.

Draw a diagram of the arrangement you use to get the clean water.

You must not drink this water. Why not?

Find out how water is cleaned at a water treatment plant.

Things to do

1 Imagine you are a water particle in a drop of rain. Write about your adventures as you pass through the water cycle. Finish your story when you reach the clouds again. Be sure to write about changes of state.

2 Draw a diagram of the apparatus used normally for filtration in the laboratory. Label the **residue** and **filtrate** on your diagram.

3 In some areas sodium fluoride (NaF) is added to the water after it has been cleaned. Dentists think that this **fluoride** helps to stop tooth decay. However too much fluoride can be poisonous.
a) Do you think fluoride should be added? Explain your views.
b) Suggest other ways of reducing the decay of your teeth.

4 Find out about **hard water**.
What is the difference between **hard water** and **soft water**?
Give one advantage and one disadvantage of living in a hard water area.

What's in a cloud?

20c

Is it a cloudy day today?
Cloud cover is the amount of sky covered by cloud.
It is measured in eighths:

a Look at the sky. What is the cloud cover today?

Clouds can have many different shapes and sizes.
The height at which they form and the speed of the wind affects
their shapes.

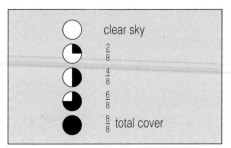

○	clear sky
◔	$\frac{2}{8}$
◑	$\frac{4}{8}$
◕	$\frac{6}{8}$
●	$\frac{8}{8}$ total cover

Cloud cover

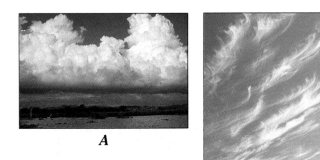

A

B

C

D

Photographs A–D show the most common types of clouds.

▶ Look at these descriptions of clouds:

Cumulus	Looks like cotton wool. It has a flat base with domes above.
Cumulonimbus	They are very tall. The top of the cloud often spreads out.
Cirrus	Feathery streaks.
Stratus	A flat cloud which is grey all over.

b Use these descriptions to identify each of the photographs A–D.
Write down your answers.

c What do the clouds outside look like today?

How are clouds made?

The Sun gives energy to the water in rivers, lakes and the sea.
This makes the water **evaporate**. The liquid water changes into a gas.
This is called **water vapour**. The water vapour rises and meets
colder air. The water vapour cools and **condenses**. Tiny droplets
of moisture form. These are called **cloud droplets**. They make up
the **clouds**.

LIQUID ⟶ EVAPORATES ⟶ GAS
⟸ CONDENSES ⟸

Thinking about clouds

- Soak a paper towel in water.

  Put the towel in a dish. Stand the dish on an electric balance.

  Record the mass every 20 seconds for 5 minutes.

  Now repeat the experiment, gently warming the wet towel with a hair dryer.

  What do your results show?

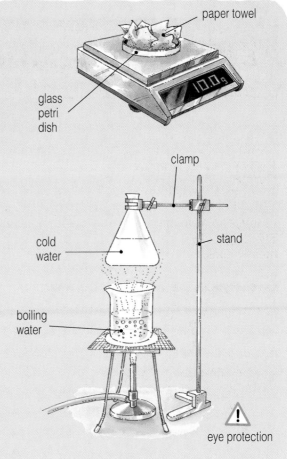

paper towel

glass petri dish

clamp

cold water

stand

boiling water

eye protection

- Half fill a conical flask with cold water.

  Clamp the flask about 25 cm above a beaker of water. Move the stand, clamp and flask to one side while you heat the beaker of water.

  When the water is boiling, **carefully** move the stand so the flask is above the beaker.

  Record what you see.

How do your experiments fit into the story of how clouds are made?
Which processes do the experiments show?
Explain what happens to the molecules of water in each experiment.

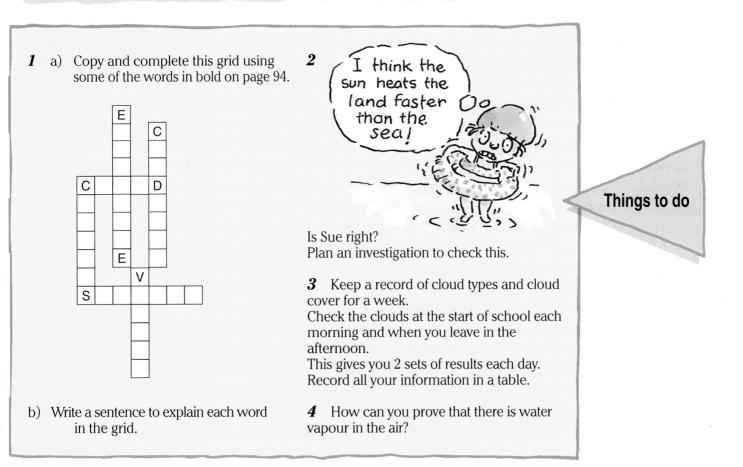

1 a) Copy and complete this grid using some of the words in bold on page 94.

E
C
C   D
E
V
S

b) Write a sentence to explain each word in the grid.

2

I think the sun heats the land faster than the sea!

Is Sue right?
Plan an investigation to check this.

3 Keep a record of cloud types and cloud cover for a week.
Check the clouds at the start of school each morning and when you leave in the afternoon.
This gives you 2 sets of results each day.
Record all your information in a table.

4 How can you prove that there is water vapour in the air?

Things to do

Why does it rain?

▶ Think about countries all over the world.
Make a list of the problems that can happen when there is:

a too much rain,

b too little rain.

Clouds and rain

▶ Copy this diagram:
Write a few lines at the side of the arrow to explain how the
cloud has formed.

Clouds are made of tiny droplets of water. This is called **condensation**.
A cloud gives rain when the cloud droplets get bigger and bigger to form
raindrops. The raindrops get so heavy that they fall to the ground.

Sometimes we think that we have more than our fair share of rain!
Britain can get rain at any time of the year. One reason for this is
that Britain is surrounded by sea. Winds that travel across the sea
pick up moisture. This means that there is a chance of rain as the
moist air cools over the land. The most common winds heading for
Britain have crossed a large ocean. They are sure to be carrying lots
of water!

▶ Use an atlas to find out the areas of land and sea which
surround Britain.

We're prepared for the British summer.

```
        N
NW   |   NE
   \  |  /
W —  🗺️  — E
   /  |  \
SW   |   SE
        S
```

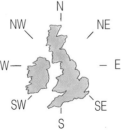

c Which winds would bring the wettest weather? Why?

d Which winds would bring the driest weather? Why?

True or False?

People love to talk about the weather! Maybe you've heard some comments like those below. Discuss them in your group. Decide whether each one is *true* or *false*. Keep a note of the reasons for your decision. Your teacher may ask you to explain your ideas.

As clouds pass over mountains they burst and let all their water out.

The risk of floods gets greater as we build more towns and cities.

The higher up you live, the less rain you'll get.

When the cows are sitting down in a field it's about to rain.

If it rains on St. Swithin's Day, it will rain for the next 40 days.

Looking at seaweed tells you about the weather.

All life on Earth needs rain.

Rainfall is involved in the process of making sedimentary rock.

Does your group agree with Ivor Grouch?

Write a letter to the newspaper to support or object to Ivor Grouch's views.

Another drought!

We get a few days of sunshine in this country and we're told there's a drought. We've got to save water. A leaflet has come through my door to tell me what I can't do.

I'm fed up with it. I don't like showers. Why can't I have a bath if I want? Why can't I water my garden? I've spent hundreds of pounds on my plants *and* I pay for my water!

A country where it rains as much as it does here should never have water shortages. Who is mismanaging the water supply?

Ivor Grouch

Extract from
The Dampshire Herald
August

Things to do

1 Copy and complete using the words in the box:

> raindrops   cools   droplets   rises

Clouds form when moist air . . . . and then . . . . so that it changes to cloud . . . . When these . . . . become bigger they form . . . . These then fall to the ground as rain.

2 Look at the design of your school. How is your school designed to stop problems of flooding?
Draw any important features.

3 Design a piece of apparatus to measure rainfall. Your apparatus should be able to remain outdoors for long periods of time.

4 Find out about 3 different types of rainfall:
a) relief rainfall
b) convectional rainfall
c) frontal rainfall.

5 Will it rain for 40 days after a rainy St. Swithin's Day? There are lots of sayings about the weather. Use books and ask friends or relatives about these sayings. Write down as many as you can. Do you think there is truth in any of these?

Questions

1   What can cause water pollution?
Make a poster to warn of the dangers of water pollution.

2   Do a survey of different bottled waters.
What substances do they contain?
How much do they cost?
Why do people buy bottled water?

3   Jo has drawn a simple water cycle.

Describe what is happening at each stage where Jo has drawn an arrow.

4   Rain is caused by moist air rising and cooling.
a)   Why does it rain a lot in the west of Britain?
b)   Why is it important to be able to predict when rain is coming?
Make a list of people who need to know about rainfall.

5   Write a poem about clouds and rain using some of the following words:

moisture   droplet   condensation   storm   flood   sea
evaporate   vapour   grey   wind   liquid   hill

6   Bangladesh has had very bad floods in recent years. The country has heavy monsoon rains. A major reason for the floods is that trees have been cleared from the hills on which the rain falls. The trees have been removed to burn as fuel and to increase the amount of farmland.
a)   Use an atlas to find Bangladesh. Name 2 rivers which are likely to overflow near Dhaka.
b)   Why does clearing the trees cause floods during the monsoon?
c)   What sort of help do the people of Bangladesh need during the floods?
d)   How could the people of Bangladesh be helped to stop floods in the future?

7   Rajid says that the rate at which water drains through a soil depends on the type of soil.
Plan an investigation to test Rajid's idea.

Energy

21

Without energy nothing can ever happen!

All living things need energy to stay alive and to move.
You get your energy from food.

Our homes, transport and factories need the energy that comes from fuels.
But the world is running out of fuel . . . .

a Name 4 things in your home that use electricity.

b Why do we need to eat food?

c Make a list of 5 things that you have done today. Put them in order, starting with the one that you think needs most energy.

Energy which is stored is called **potential** energy. An object which is moving has **kinetic** energy.

d Complete this sentence:
When a clockwork toy is moving, ... energy in the spring is ***transferred*** to . . . . energy.

▶ Look at the diagram. It shows an electric motor lifting up a weight:

e What kind of energy does the moving wheel have?

f Where does this energy come from?

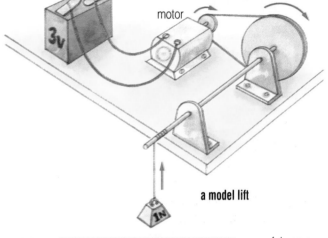

battery

motor

a model lift

g Copy and complete its **Energy Transfer Diagram**:

. . . . energy in the battery **10 J**

useful energy lifting up the . . . . **7 J**

wasted energy heating up the . . . . **3 J**

h From the diagram, what can you say about the amount of energy (in joules) ***before*** the transfer and ***after*** the transfer?

i How much of the energy after the transfer is useful? What has happened to the rest?

We say its **efficiency** is 70%, because only 7 out of 10 joules have done something useful.

This is what usually happens in energy transfers. Although there is the same amount of energy afterwards, not all of it is useful.

▶ Now look at this diagram:

j What happens to the lamp? Why?

k Suppose the weight starts with 100 J of potential energy, and then 20 joules appear as light energy shining from the lamp. What has happened to the other 80 joules?

l What is the efficiency in this case?

m Draw an Energy Transfer Diagram for this, and then label it.

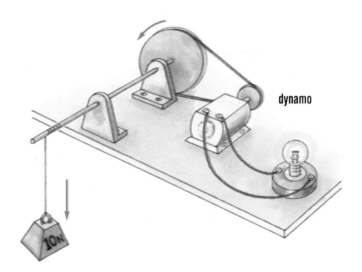

dynamo

Energy transfers

Your teacher will show you some of these.
Observe them carefully, and think about the energy transfers.
For each one, draw an Energy Transfer Diagram and label it.

n   an electric kettle

o   a clockwork toy

p   a solar-powered
        calculator

q   a signal-generator
        and loudspeaker

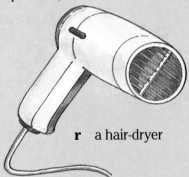

r   a hair-dryer

s   a steam engine and dynamo

Things to do

1   Copy and complete:
a)   Energy is measured in . . . .
b)   Stored energy is called . . . . energy.
c)   A moving object has . . . . energy.
d)   In any energy transfer, the total amount
     of . . . . before the transfer is always . . . .
     to the total amount of . . . . afterwards.
e)   After the transfer, not all of the . . . . is
     useful.

2   Describe the energy transfers involved in:
a)   a torch,
b)   a TV set,
c)   playing a guitar.

3   What happens if the food you eat
contains more energy than you need?

4   Write down 3 examples in everyday life
where potential energy is transferred to
kinetic energy.

5   Are some kettles cheaper to run than
others? Plan an investigation to compare
some kettles.
How will you make it a fair test?

6   Why do you think it is impossible for
anyone to build a perpetual-motion machine?

Energy from the Sun

a Where does the Earth get most of its energy from?
b Think of all the things that happen because of the Sun.
   Make a list of as many as you can.

Making food

Green plants can capture the energy in the sunlight.
The green chemical in their leaves is called **chlorophyll**.
It **absorbs** the Sun's energy and uses it to make food.
It also makes oxygen for us to breathe.
This process is called **photo-synthesis**.

Because plants make food, they are called producers.
Animals eat this food – they are consumers.

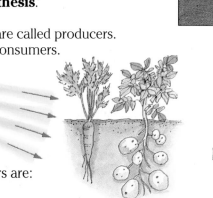

In this picture, the energy transfers are:

Sun ➡ vegetables ➡ human

This is called a **food-energy chain**.

Here are some food-energy chains that are in the
wrong order. Write each one in the correct order.
c rabbit, Sun, grass, fox
d humans, grass, sheep, Sun
e thrush, Sun, cabbage, caterpillar

Biofuel

Plant and animal materials are called **biomass**.
As well as being food, biomass can give us energy in other ways:

* Wood is a fuel. It can be burnt to give energy for heating.

* In Brazil they grow sugar cane, and then use the sugar to make
  alcohol. The alcohol is then used in cars, instead of petrol.

* Rotting plants and animal manure can make a gas called methane.
  This is like the gas you use in a Bunsen burner.
  If the plants rots in a closed tank, called a **digester**, the gas can be
  piped away. This is often used in China and India:

> ▶ Design a digester to use straw and dung. Think about:
>   * It needs an air-tight tank.
>   * How will you get the gas out, to a cooker?
>   * How will you get the straw and dung in?
> **f** Draw a sketch of your design and label it.

A cow-dung digester in India

Solar energy

The energy in the Sun's rays is called **solar energy**.

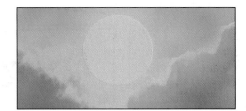

g Why is the Earth the only planet in the Solar System with life on it?

h Which parts of the Earth get the least energy?

▶ Look at these 3 ways of using solar energy, and answer the questions:

A **solar cell** transfers some of the sunlight into electricity.

In the photo, some solar cells are being used to run an electric water-pump:

You may have a calculator that uses a solar cell.

i What are the advantages and disadvantages of a solar-powered calculator?

j Why are solar cells not widely used?

A **solar cooker** has a curved mirror, to focus the Sun's rays:

k Is the mirror convex or concave?

▶ Design your own solar cooker. Think about:
- Should the mirror be large or small?
- Should the mirror be fixed or adjustable?
- Where should you put the pan or kettle?

l Draw a sketch of your design and label it.

Some houses have a **solar panel** on the roof.
The water in the panel is heated by the Sun, and stored in a tank:

▶ Design your own solar panel system. Think about:
- Hot water rises, cold water falls.
- Black cars get hotter in the Sun than white cars.
- Objects get hotter behind glass (like in a greenhouse).

m Draw a sketch of your design and label it.

Things to do

1 Copy and complete:
a) Energy from the Sun is called .... energy.
b) Green plants contain a chemical called .... This absorbs the Sun's .... and uses it to make .... and .... This process is called ....
c) The materials that plants and animals are made from, are called ....

2 How does a greenhouse use solar energy to help gardeners?

3 For each of these food-energy chains, write them in the correct order:
a) chicken, Sun, human, corn
b) seaweed, Sun, seagull, snail
c) ladybird, rose, greenfly, Sun
d) grasshopper, lizard, Sun, grass, hawk
e) bee, human, flower, Sun, honey
f) dead leaves, frog, Sun, earthworm, tree

4 What are:
a) the advantages, and
b) the disadvantages of a solar panel?

From fossils to fuels

A fossil in coal

▶ What is a **fuel?**
Write down as many fuels as you can think of.

▶ Coal, oil and natural gas are important fuels. Read the sections below and then answer questions **a** to **j**.

How was coal formed?

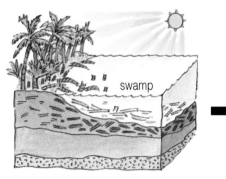

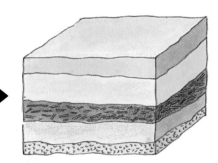

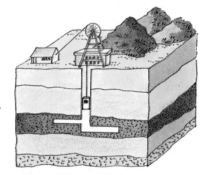

300 million years ago, plants store the Sun's energy. Dead plants fall into swampy water. The mud stops them from rotting away.

As the mud piles up, it squashes the plants.
After millions of years under pressure, the mud becomes rock and the plants become **coal**.

To reach the coal, miners dig shafts and tunnels. There is probably enough coal to last 300 years.
Fossils of plants are sometimes found in lumps of coal.

How was oil formed?

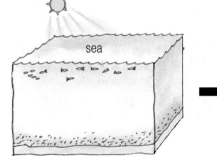

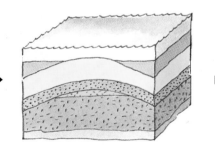

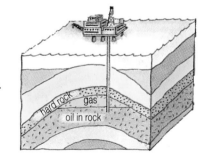

Tiny animals live in the sea.
When they die, they fall into the mud and sand at the bottom, and don't rot away.

Over millions of years they get buried deeper by the mud and sand.
The pressure changes the mud and sand into rock, and the dead animals become **crude oil** and **natural gas**.

The oil can move upwards through some rocks, but if it meets a layer of hard rock it is trapped (with the gas).
An oil rig can drill down to release it.
There is enough oil to last about 40 years.

a What is a fossil?

b Why are coal, oil and gas called fossil fuels?

c Explain in your own words how coal was formed.

d Give 2 similarities and 2 differences between the way coal was formed and the way oil was formed.

e Diana says, "The energy stored in coal, oil and gas all comes from the Sun." Explain this statement.

f Oil, coal and gas are called **non-renewable** resources. What do you think this means?

g Why will fossil fuels eventually run out?

h How old will you be when the oil runs out?

i Why is coal usually found in layers?

j Why are some rocks called sedimentary rocks?

Renewable and non-renewable

Some sources of energy are **renewable**.
For example, wood. It can be burnt, but a new tree can be planted.
Solar energy is also a renewable resource.

However coal, oil and gas are **non-renewable** resources. Once we have used them up, they are gone forever.

Uranium is another non-renewable resource. It is used in nuclear power stations. Supplies of uranium will run out eventually.

Making electricity

The graphs show the sources of energy used to generate electricity in 3 countries:

k Which country generates the most electricity?

l List the energy sources used to make electricity in the UK.

m Which is the smallest source of energy in the UK? Why is this?

n Which is the main source in Norway? Why is this?

o What do you notice about the use of nuclear power in the 3 countries?

p Which of the sources is renewable?

q Coal has been used a lot in the UK, but coal-fired power stations can produce a lot of **acid rain**. What would be your solution to this problem, taking into account:
i) the environment?
ii) coal-miners and their families?

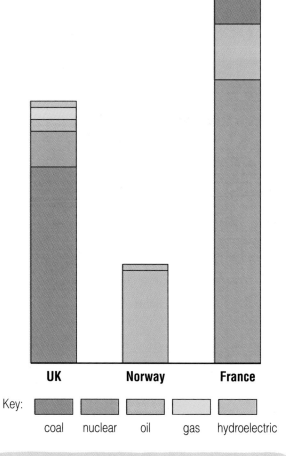

UK   Norway   France

Key: coal nuclear oil gas hydroelectric

The nuclear debate

The world is running out of energy, but many people are against the use of nuclear energy.

In a group, use the Help Sheets to discuss your ideas *for* or *against* using nuclear energy.

Things to do

1 Copy and complete:
a) Coal, oil and . . . . are fossil . . . .
They have taken . . . . of years to form.
b) Coal, oil, gas and uranium are non- . . . .
sources of energy.
c) Some renewable sources of energy are:

2 Explain in your own words how oil and natural gas were formed.

3 Why is it important to avoid wastage of fossil fuels?

4 Make a table to show how people could save non-renewable fuels. For example:

Action to be taken	How it saves fuel
don't overfill the kettle	uses less electricity

Burning fuels

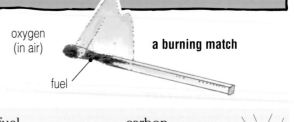

oxygen (in air)

fuel

a burning match

▶ What fuels do you use in your home? Make a list.

▶ Look at this picture of a *match* burning:

Wood is a fuel. It has potential energy stored in it.
This energy can only be transferred if the fuel burns with
the oxygen in the air.
It is a chemical reaction.

$$\text{fuel (wood)} + \text{oxygen} \rightarrow \text{carbon dioxide} + \text{water} + \text{energy}$$

Burning fuels are used in power stations and in car engines.

As you know from topic 18, the *cells* in your body 'burn' the
food that you eat – but of course there aren't any flames!
Your blood carries sugar (from your food) and oxygen
(from the air you breathe) to all the cells in your body:

The *same* chemical reaction releases the energy.

This is called **respiration**.
You use the energy to keep warm and to move around.

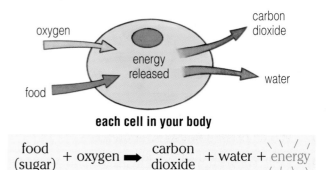

oxygen

food

energy released

carbon dioxide

water

each cell in your body

$$\text{food (sugar)} + \text{oxygen} \rightarrow \text{carbon dioxide} + \text{water} + \text{energy}$$

Using fuels in a power station

Follow the diagram to see how fuel is burned, to generate electricity:

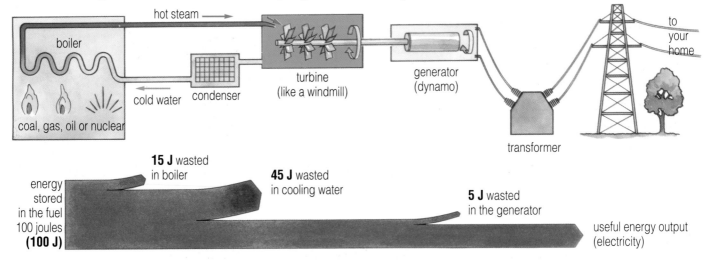

hot steam →

boiler

cold water   condenser

turbine (like a windmill)

generator (dynamo)

coal, gas, oil or nuclear

transformer

to your home

15 J wasted
in boiler

45 J wasted
in cooling water

5 J wasted
in the generator

energy
stored
in the fuel
100 joules
(100 J)

useful energy output
(electricity)

a   What does the boiler do?

b   What does the steam do to the turbine?

c   What does the generator do?

d   For every 100 joules of energy in the fuel, how much comes out as useful energy?

e   What is the **efficiency** of the power station?

f   Where is most energy wasted? Can you think of a way of using this wasted energy?

Comparing engines

A car engine burns fuel. The energy from the fuel makes the engine turn, so the car moves.

A human body is also a kind of engine:

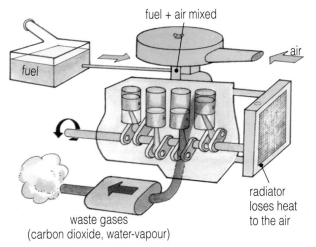

fuel + air mixed

air

fuel

waste gases
(carbon dioxide, water-vapour)

radiator
loses heat
to the air

air

food

waste gases
(carbon dioxide,
water-vapour)

blood carries
sugar (from food) and
oxygen (from air) to
each cell,
it also takes away
the waste products

in the cells,
energy is released

unused
parts
of food

water

skin sweats
to lose heat

For each question **g** to **m**, write down the answer, first for the *car engine* and then for the *human engine*.

g What fuel does the engine use?

h What does it use the fuel for?

i Where is the fuel used ('burned') in the engine?

j How does the engine get its oxygen?

k How does the engine get rid of unwanted heat?

l What waste substances are produced in the engine?

m How does the engine get rid of these waste products?

Investigating sweeteners

Plan an investigation to compare the *amount of energy* in some *sugar* and in some *artificial sweetener*.

- How will you make them burn?
- What will you do with the energy from the burning fuel?
- How will you make it a fair test? And safe!

Ask your teacher to check your plan, and then do the investigation.

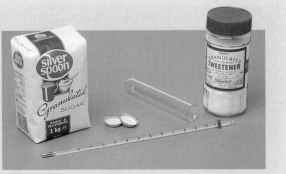

1 Copy and complete:
a) In order to burn, a fuel needs . . . . It usually gets this from the . . . .
b) The chemical equation for a fuel burning is:
c) In the cells in my body, energy is released when sugar (from my . . . .) reacts with . . . . (from the . . . . I breathe).
d) This is called . . . .
e) The chemical equation for this is:

2 Do a flow-chart to show how the energy in a coal-mine makes a cup of tea for you.

3 Draw an Energy Transfer Diagram for:
a) a burning match,
b) a cell in your body.

4 Explain how you should deal with:
a) a chip-pan fire, b) a petrol fire,
c) an electric blanket on fire.

Things to do

Energy for ever?

▶ What is meant by a **non-renewable** resource?
Name 4 non-renewable sources of energy.

What is a **renewable** source of energy?
You have already studied 2 renewable sources,
biofuel and **solar energy** (on pages 102–103).
Here are some more:

Wind energy
Windmills have been used for centuries.
Modern wind-turbines are huge. One advantage is that the wind blows most when we need more energy – in winter.

Why does this energy come originally from the Sun?

Geothermal energy
This geyser is spurting out hot water. This is because deep inside the Earth it is very hot.
If a very deep hole is drilled, cold water can be piped down, to return as hot water.

Wave energy
Waves are caused by the wind. They contain a lot of energy but it is hard to make use of it.
One idea is to have floats that move up and down with the waves and so turn a generator.

Why does this energy come originally from the Sun?

Hydro-electric energy
Water stored by a dam has potential energy.
When it runs down-hill, its kinetic energy can turn a turbine or a water-wheel. This can turn a generator to make electricity.

Why does the energy come originally from the Sun?
Why is this resource impossible in some countries?

Tidal energy
The tides are caused by the pull of the Moon and Sun.
In some places there are very high tides.
The water can be trapped behind a barrier, like a dam:
Then it can be used like hydro-electric energy.

Joule Island

Joule Island is a remote island, in the Pacific Ocean.
You are in a team of 30 scientists who will be staying on the island for 3 years to study it.
Your task is to provide all the energy that the team will need.

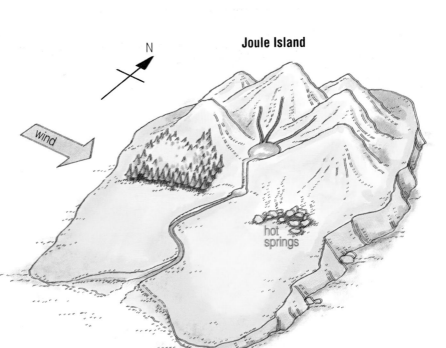

Joule Island

Study the island:

a There are no fossil fuels on the island, and it is 500 km to the mainland. What would be the advantages and disadvantages of setting up a power station which used coal or oil?

b What renewable energy resources are there on the island?

c Which natural resource on the island should be carefully conserved?

The island has sunny days but cold nights.
The wind blows most days, but not in summer.
The hot springs are at a temperature of 80°C.

d The team is going to build huts to live in. Name 2 ways in which the huts could be heated.

e Design a way of supplying hot water for washing.

f Design a way of supplying energy for cooking food.

g The team has a refrigerator for medicines which have to be kept cool at all times, day and night. Design a way of supplying electricity continuously for the refrigerator.

h Your teacher will give you a picture of the island. Mark on the picture:
 • where you would build the huts, and
 • where you would build any energy installations you have designed, and
 • show how the energy would be transferred to the huts.

i When summer comes, you find that the fresh-water sources tend to dry up. How does this affect your energy plan? Design a way to get over this.

j For some experiments on the island you need some gas for a Bunsen burner. Describe 2 ways to provide this.

---

1 Copy and complete:
a) Fossil fuels like coal, ... , and natural gas will eventually run out. They are non-. . . . resources.
b) Nuclear energy is also a . . . . resource.
c) Other energy resources are called . . . .
d) There are 7 renewable sources of energy. They are:

2 Which of the 7 renewable sources get their energy originally from the Sun?

3 Design the scientists' huts for Joule Island. You should design them so that:
 • they have sleeping, leisure and work areas, and
 • they will be cool during the day and warm at night.

Things to do

Questions

1   Draw Energy Transfer Diagrams for:
   a) a torch
   b) a bonfire burning
   c) a boy kicking a football.

2   A car engine is only 25% efficient. Of every 100 joules in the petrol, only 25 J actually make the car move.
   a) What happens to the other 75 joules?
   b) Draw an Energy Transfer Diagram for the car.

3   Draw a food-energy chain to show how the energy in a cheeseburger comes from the Sun to:
   a) the cheese
   b) the bread.

4   In a solar cell, for every 100 joules of solar energy shining on it, only 10 J is transferred to useful energy in electricity.
   a) What happens to the other 90 joules?
   b) What is the efficiency of the solar cell?
   c) Draw an Energy Transfer Diagram of this, with the width of the arrows to scale. Label it.

5   Think about what life would be like without coal, oil, or natural gas. (Remember that petrol and plastics come from oil.)
   You can present your ideas in a list, or in a story, or on a poster.

6   When you switch on a light, it is the result of a long chain of events. These are listed below, in the wrong order. Write them down in the correct order.
   A   plants take in energy from the Sun
   B   coal is burnt in oxygen (in air)
   C   water is heated, to make steam
   D   the Sun produces energy
   E   plants change to coal over millions of years
   F   steam makes a turbine turn
   G   the generator produces electricity
   H   electricity heats up the lamp and it shines
   I   the turbine turns a generator
   J   electricity travels through wires to your home

7   A typical British family uses energy like this:

Heating the house	40%	Heating water	10%
Transport	25%	Food eaten	5%
Electrical goods	15%	Cooking	5%

   a) Draw a pie-chart or a bar-chart to show this information.
   b) Where should they look first in order to save money?

Have you seen fireworks that look as good as this?
Each flash of light means that a chemical reaction is happening.

Some reactions are much slower.
They seem less exciting.
But *all* chemical reactions are important
. . . they are keeping you alive today!

Scientists – the makers!

22a

What can you remember about crude oil?
Do you remember that it can be separated into different substances?

▶ Make a list of some of the substances we get from crude oil.
Say what each substance is used for.
(If you need a clue, look back to page 30.)

Getting useful substances from crude oil is an example of a
manufacturing process.

raw material	*manufacturing process*	*useful product*
e.g. crude oil		e.g. petrol

One of the substances in crude oil is **naphtha**. It is very useful
because it can be made into lots of other things.
Paints, medicines and fibres can all be made from naphtha.

Making new materials is a very important job for a scientist.
▶ Look at the picture below. What materials have scientists
helped to make? Write a list. For example, glass for windows.

Do you remember burning magnesium ribbon?
This was an example of making a new substance.

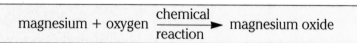

A change which makes a new substance is called a **chemical change**.
A **chemical reaction** must happen.
The new substance is called a **product**.

$$\text{magnesium} + \text{oxygen} \xrightarrow[\text{reaction}]{\text{chemical}} \text{magnesium oxide}$$

a What is the product of this reaction?

b How do you know a new substance has been made?

▶ Which of **c** to **g** are chemical changes?
(Hint: Can you get the materials you started with back again
easily?
or Have new substances been made?)

c Baking a cake.   **f** Dissolving salt in water.
d Striking a match.   **g** Burning some toast.
e Making ice cubes.

112

Notice any change?

Here are some tests for you to do. In each one find out what happens when you mix the substances.

For each test, say if you think a new substance is made. Before you start, think about how to record your findings.

- You must wear safety glasses.
- Only use small amounts of the substances.
  Do not use more than the instructions tell you to.

⚠ eye protection

1 3 cm^3 dilute sulphuric acid + $\frac{1}{2}$ spatula measure of copper carbonate
2 3 cm^3 dilute sulphuric acid + 3 cm^3 sodium hydroxide solution
3 3 cm^3 vinegar + $\frac{1}{2}$ spatula measure of bicarbonate of soda
4 3 cm^3 water + $\frac{1}{2}$ spatula measure of copper oxide
5 3 cm^3 lead nitrate solution + 3 cm^3 potassium iodide solution
6 3 cm^3 dilute sulphuric acid + copper foil
7 3 cm^3 water + iron nail
8 3 cm^3 dilute sulphuric acid + 2 cm of magnesium ribbon
9 3 cm^3 copper sulphate solution + 1 spatula measure of iron filings

Look at your results for all the tests.
How do you know if a new substance has been made?

h Make a list of things that can happen when a new substance is made.
These things can tell you that there has been a **chemical reaction**.

---

1 Copy and complete:
a) A change which makes new substances is called a c . . . . change.
b) This change is called a c . . . . r . . . .
c) The new substance is called a p . . . .
d) A p . . . . of the r . . . . between carbon and oxygen is called c . . . . d . . . .

2 Make a list of some chemical changes that happen around your home.
Draw a picture to show one of the changes happening.

3 Do a survey of a car.
Make a list of all the materials used to make the car.
Say whether these are natural or made materials.
Why is each material right for its job?

Material	Natural? or Made?	Why is it used?

4 Think of a raw material which is made into a useful substance.
Draw a poster to show this change in an interesting way.

Things to do

Getting the iron

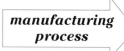

Look at the photos above. What do all the objects have in common? There's more than one answer to this! But if you've read the top of the page, you've probably said they all contain *iron*. You're right!

Look at the photo of the rock. It is iron **ore**.
It comes from the ground. The ore itself isn't useful. But it can be made into other useful materials.

raw material	*manufacturing process*	*useful products*
iron ore	→	iron and steel

▶ Why do you think iron is so useful?

a Make a list of the properties of iron.

b What is steel? Does it have any advantages over iron?

Every day the chemical industry makes many useful products from raw materials. This can involve lots of **chemical reactions**.
Let's look at the reactions needed to make iron.

The blast furnace

The most common ore of iron is called haematite. It is iron oxide. This is a compound of iron and oxygen. To get the iron from this ore, we need to remove the oxygen. This is done in a **blast furnace**:

$$\text{iron oxide} \xrightarrow{\text{remove the oxygen}} \text{iron}$$

We say the iron oxide is **reduced**.
Reducing means *taking the oxygen away*.

In the blast furnace there is some coke, a form of carbon. This helps to take the oxygen away. It does this at a temperature of 1200°C.
But there are lots of impurities in the iron ore.
Limestone is used to get rid of these.
In the heat, limestone breaks down. It changes to new substances.

$$\text{calcium carbonate} \atop \text{(limestone)} \xrightarrow{\text{heat}} \text{calcium oxide} + \text{carbon dioxide}$$

The calcium oxide then reacts with some impure substances in the iron ore.
It's very hot inside the blast furnace so the iron made is melted (molten).

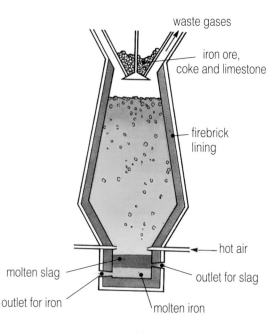

waste gases
iron ore, coke and limestone
firebrick lining
hot air
molten slag
outlet for slag
outlet for iron
molten iron

c Name 3 substances used in the blast furnace.

d Why is the blast furnace lined with fire brick?

e Name one gas which will be a 'waste gas'.

f Why do blast furnaces work every day and night, all year round?

g Why is iron the cheapest of all metals?

The size of the van in this photograph shows you the blast furnace is very big.

Heating limestone

One of the reactions in the blast furnace involves heating limestone (calcium carbonate). You can see what happens for yourself.

Before you start to heat:

* Ask your teacher to check your apparatus.
* Make sure you know how to stop suck-back (remove the lime water before you stop heating the limestone).
* Think about why you use lime water in the experiment.

Heat the limestone gently at first, then more strongly.

h What happens?

i Try to write a word equation for this chemical change.

Use this experiment to help you plan an investigation on some other carbonates.

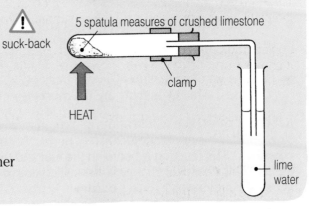

suck-back

5 spatula measures of crushed limestone

clamp

HEAT

lime water

Investigation

Which carbonate changes the fastest?
Your teacher must check your plan.
If there is time you may be able to carry out the investigation.

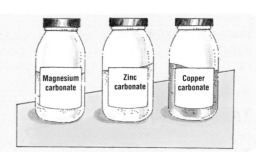

Magnesium carbonate   Zinc carbonate   Copper carbonate

---

1 Copy and complete using the words in the box:

manufacturing	furnace	oxygen
raw	coke	impurities	hot
ore	carbonate	product

A .... process changes a .... material into a useful .... Iron .... is made into iron in the blast ....
The iron ore is reduced. This means its .... is removed. .... is a form of carbon which helps to reduce iron ore. Limestone is calcium .... It breaks down in the heat. It is used to get rid of .... The iron made is melted because it is so .... in the furnace.

2 The words in the box are ores.

| bauxite | galena |
| malachite | magnetite |

a) Use books to find out which metal comes from each ore.

b) Write down one use of each metal.

c) What things affect the price of any metal?

d) Iron is quite cheap. Name 2 expensive metals.

3 Limestone is an example of a **sedimentary** rock.

a) Explain how this type of rock forms.

b) Name 2 other types of rock. Write a few lines to explain how each one forms.

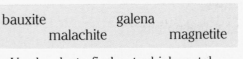

Things to do

A burning problem

Have you ever seen this symbol:
This is called the **fire triangle**.

a Why are the words HEAT OXYGEN and FUEL written on the fire triangle?

b Use the fire triangle to explain what you must do to stop a fire.

You have already learnt about some fuels in your science lessons.

▶ Make a list of all the fuels you can think of.
Choose 3 which you think are the most important. Explain why.

A fuel is a substance which burns in oxygen to give us energy. New substances are made as the fuel burns. Burning is a **chemical reaction**. The reaction is between the fuel and the oxygen gas in the air. This chemical change is also called **combustion**. When the fuel burns it makes **oxides**.

fuel + oxygen ⟶ oxides + energy

This is an **exothermic** reaction. This means heat (energy) is given out.

Burning a fuel

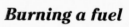

Bread is an example of a fuel. It is fuel for your body. Usually you don't want to burn the toast. But in the next experiment you will do just that!

Hold the bread in tongs over a heat-resistant mat. Observe the bread very carefully. Heat the bread until it no longer burns. Write down everything you see.

⚠ eye protection

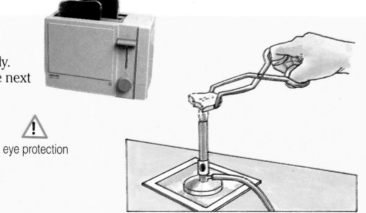

What happens to the bread during burning?

In your group, discuss your ideas about burning.

Bread is like most fuels. It contains hydrogen and carbon.

That must mean it makes oxides of carbon and hydrogen when it burns.

Where do the oxides go?

c When carbon reacts with oxygen it makes . . . .

d When hydrogen reacts with oxygen it makes . . . . This is usually called . . . .

e How could you test the substances made in **c** and **d**?

Burning candles . . . what do *you* think?

A candle is made of wax. Candle wax is a fuel. It contains the elements **carbon** and **hydrogen**.

Predict what will happen when a beaker is put over a burning candle.
What do you expect to see**?**

Now try the experiment. Were you right**?**

⚠️ eye protection

A more difficult prediction . . .
Predict what will happen when a beaker is put over candles of three different heights.
Which candle do you think will go out first**?**
Explain your ideas.

Now try the experiment. Were you right**?**
Try to explain what happened.

candle wax + oxygen ⟶ carbon dioxide + water + energy

This reaction is an example of an **oxidation**. A substance has **gained** oxygen.

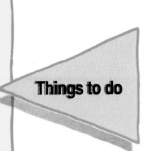

Things to do

1   Write 2 or 3 lines about each of the following words to explain what it means.
a) fuel
b) combustion
c) oxide
d) exothermic

2   You need to take great care when using fuels. Draw a poster to warn people of the dangers.

3   Ethanol is a substance which is made from carbon, hydrogen and oxygen. Copy and complete the word equation to show what you think happens when it burns.

ethanol + ... ⟶ ... . + water + ...

4   Collect some newspaper cuttings about fires. For each fire try to find out:
a) How the fire started.
b) Whether the fire could have been prevented.
c) How the fire was put out.
Record all the information in a table.

5   Fuels can be solids, liquids or gases.
Give one example of each type.
What are the advantages of
a) a **solid** fuel?
b) a **liquid** fuel?
c) a **gaseous** fuel?

In topic 18 you were 'Staying alive'.
What can you remember about **respiration?**

Every time you breathe, a chemical reaction is taking place.

a What happens when Jane breathes out into the lime water?

b What happens when Tom breathes out onto the cold window?

You get energy out of your food in respiration.

c Complete the word equation for respiration:

sugar + oxygen ———————> ... ... + ... + energy

Respiration is an **exothermic** reaction.

d What does exothermic mean?

e Think about respiration and burning.
What are the similarities between these reactions?
Make a list of your ideas.

f Why do you get short of breath and hot in a race?

Fermentation

When *we* respire we use oxygen. It reacts with our food to make carbon dioxide, water and energy.
But some living things can respire *without* oxygen.
An example is a microbe called **yeast**.
Yeast is used to make wine. It uses up the sugar in fruit to make alcohol.
This process is called **fermentation**.

Drink	Made from
wine	grapes
brandy	grapes
beer	barley
whisky	barley
vodka	potato
cider	apples
saki	rice

Bubbling alcohol

Try your own fermentation using sugar.
Mix some yeast with about 4 g of sugar.
Add about 10 cm³ of warm water (about 35°C) and put it in a flask.
Leave the apparatus set up like the diagram shows.
Look at the flask after 10 minutes. What do you notice?
Look at the flask again after 45 minutes.

⚠ **Carefully** smell the contents. What do you notice?

⚠ **Do not attempt to drink this substance. It is very impure.**

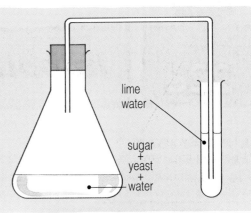

lime water

sugar + yeast + water

The word equation for fermentation is

$$\text{sugar (with yeast)} \longrightarrow \text{alcohol} + \text{carbon dioxide} + \text{energy}$$

g In what ways is this reaction **like** your respiration?

h In what ways is this reaction **different** to your respiration?

i Why is it dangerous to drink and drive?

j Some people think it's wrong to drink **any** alcohol. Do you agree? Discuss this in your group.

▶ Don't drink and drive.
Design your own poster to get this message across.

WHAT A BEAUTIFUL DAY TO DIE.

Drinking drivers cause more deaths in summer than at any other time of the year.

DRINKING AND DRIVING WRECKS LIVES.

---

1 Copy and complete:
a) When we breathe, the products are . . . . , ... and . . . .
b) The products of fermentation are . . . . , ... and . . . .
c) An exothermic reaction is one in which heat . . . .

2 Remind yourself of the 'Bubbling alcohol' experiment above.
How do you know a reaction has taken place?
List your ideas.

3 Yeast is also used in bread-making.
Find out how bread is made.
Why is yeast used?
Why could this be called a **bubbling reaction**?

4 Breath tests can be used to see if drivers have been drinking too much alcohol.

a) Make a list of arguments **for** random breath-testing (testing any driver at any time).
b) Make a list of arguments **against** random breath-testing.

Things to do

Rusting parts

Do you have a bike?
Does anyone in your family have a car?
If so, you probably know about the problem of rust!
Every year rust causes millions of pounds worth of damage.

▶ Make a list of 4 problems caused by rust.

Many companies spend lots of money trying to stop rusting.
To know how to stop rusting, we must know what causes it.

Kris and Becky investigated the conditions needed for an iron nail
to rust.
They set up 3 test-tubes and left them for a few days.

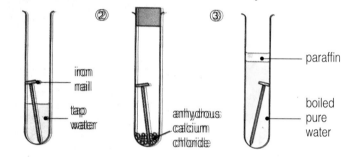

▶ Copy the table. This shows the conditions inside the tubes.

Condition	Tube ①	Tube ②	Tube ③
air	✔	✔	✘
water	✔	✘	✔

Look at the conditions in tube ②.

a What do you think anhydrous calcium chloride does?

Look at the conditions in tube ③.

b What do you think happens when the water boils?

c Why is paraffin put on top of the boiled water?

After a few days, Kris and Becky checked the tubes.

d What 2 substances must be present for iron to rust?

Have you noticed any rusting at the seaside?

e Rusting happens faster by the sea. Why do you think this is?

Rusting is an **oxidation** reaction. The iron is **oxidised**.
This means it reacts with oxygen in the air.

f Complete the word equation:

iron + oxygen ⟶ ....
(with water)

g What is the chemical name for rust?

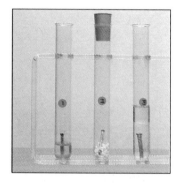

. . . after a few days

Investigating rusting

The teacher found the group's results very interesting.
She wanted Kris and Becky to do a more detailed investigation.
She didn't give them any firm ideas, but set them a question:

Imagine you are Kris. Plan an investigation into the rusting of iron.
(If you need help with this, ask your teacher for a clue!)

Stopping the rot!

To stop iron from rusting we must protect it from air and water.
Here are some ways of doing this:

- **Painting**
  This is used to protect cars and large structures such as bridges.

- **Greasing or oiling**
  This is used on moving parts of machines.

- **Plating**
  This is a thin coating of another metal which doesn't rust.
  Chromium plating is common. It gives an attractive, shiny effect.
  Zinc plating is also used. If the metal coating on the iron is zinc,
  the iron is said to be **galvanised**. Zinc will protect the iron even
  if it gets scratched.

Which rust prevention method would you use to:

h protect a lawnmower in the winter?

i protect a car bumper?

j protect a school's iron gates?

Explain your choice in each case.

Things to do

1 Copy and complete:
a) .... and .... are needed for iron to rust.
b) .... speeds up the rusting of iron.
c) The chemical name for rust is ....
d) Two methods to prevent rusting are .... and ....

2 Carry out a survey of cars near where you live. **Look** at the cars, don't **touch**. **Take care**. ⚠
Do certain makes of cars rust faster than others?
Do some parts of the car rust faster than other parts?
Present all your survey results in a table.
Write a summary report of your findings.

3 Different parts of a bike are prevented from rusting in different ways.
Make a table to show the different bike parts and the rusting prevention methods.

Parts of a bike	How is rusting prevented?
handlebars	

4 Do you think rusting happens to the same extent in all parts of the world?
Explain your ideas about this.

Fast or slow?

Do you think that rusting is a **fast** or **slow** reaction?

Chemical reactions happen at different rates.
Some are slow. Others are very fast.

▶ Think about the reactions below.
Say whether you think they are fast or slow reactions.

a Dynamite exploding.

b Wine fermenting.

c Milk turning sour.

d A sofa burning.

e Rocks being weathered.

f Magnesium burning in air.

g Developing a photograph.

h Wood burning.

i Baking a cake.

Wouldn't it be useful if we could change the rate of some reactions?

▶ Make a list of reactions which people would like to slow down.

Make a list of reactions which people would like to speed up.

Compare your answers with those of others in your group.

Do you all agree?

Chemical magic

Scientists can make reactions happen at different rates.
You may be impressed by your teacher's magic!
Your teacher will show you a reaction which makes a substance
called **iodine**.
It is easy to see when you've made iodine.
(You might remember the starch test from topic 16.)
When iodine is added to starch solution, it turns deep blue.
Your teacher will be able to make this blue colour appear at
different times . . . But you'll need to watch carefully.

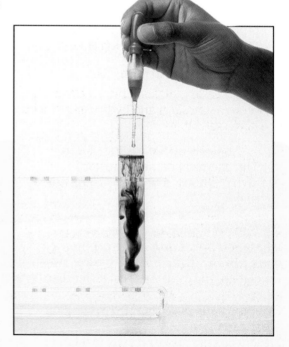

j Why do you think this reaction is called the **iodine clock**?

k How does your teacher do this? Is it **really** magic?

Speed it up!

You can investigate the reaction between magnesium and dilute hydrochloric acid.
How could this reaction be made faster or slower?
What is your hypothesis? Do you have more than one idea to test?

Your teacher will give you some magnesium and some bottles of acid for the tests.

 eye protection

- Write down the tests you would like to do.
- You must write down the equipment you need.
- You must say **exactly** how much of each of the substances you need. Do this for each test.

Let your teacher see your plans. If they are safe, you will be given the magnesium and acid. You can then start the investigation.

- Write down all the results from your tests.
- Look carefully at your results. Write a conclusion for your investigation. Your conclusion should start like this:

The reaction rate can be changed by changing ....
....
....

These are the **variables** which change the reaction rate.

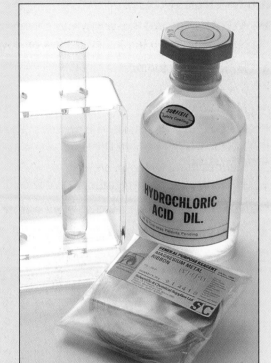

---

1 Write down 3 things which change the rate of a chemical reaction.

2 Sally was investigating the reaction between magnesium and hydrochloric acid. She had 5 sets of results.

Experiment	1	2	3	4	5
Time to collect 10 cm³ gas (in seconds)	15	6	43	15	29

a) Which experiment had the slowest rate?
b) Which experiment had the fastest rate?
c) In which 2 experiments were all the variables the same?
d) Sally says that the only thing she changed in the experiments was the temperature.
Which experiment do you think took place at the highest temperature?

3 Magnesium reacts quickly with dilute hydrochloric acid.
Look at this list of other metals:

iron   copper   zinc   calcium

a) Which of these metals would react fastest with dilute hydrochloric acid?
b) Which metal will not react with dilute hydrochloric acid?
c) Imagine you have just discovered a new metal called Spotlight.
Use the metal + acid reaction to plan an investigation to place Spotlight in the correct position in the Reactivity Series. (Hint: see pages 36–39.)

Things to do

22g A race against time!

► Look at these photographs. They show a gargoyle at Lincoln Cathedral. They were taken 100 years apart.

a Which photo was taken most recently?

b What has caused the gargoyle to change over the years?

Lots of buildings are made of limestone. The chemical name for limestone is **calcium carbonate**. It reacts with acid to make carbon dioxide.

c How can you test for carbon dioxide gas?

Carbon dioxide can be collected **over water**.

d Draw a diagram to show how you can collect a gas over water.

e Limestone reacts with the acid in rain. Why do you think it is still used to make buildings?

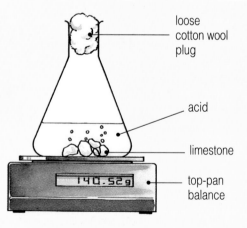

Watching an acid attack!

Your teacher will show you limestone and acid reacting.
But before the experiment starts, make a prediction.
Do you think the mass of the flask and its contents will change during the reaction?
Will the mass . . . . stay the same?
         . . . go up?
         . . . go down?
Predict what you think will happen. Why?

Draw out a results table.
You will need to record the mass every 15 seconds.

Your teacher will now start the experiment.

f What happens to the mass of the flask and its contents?

g Why do you think this happens?

h Is this a fast or slow reaction?

i Why do you think the reaction at Lincoln Cathedral is slower?

j Why do we use a loose cotton wool plug in the experiment?

loose cotton wool plug

acid

limestone

140.52g

top-pan balance

Time (seconds)	Mass of flask (grams)
0	
15	

Time it to perfection

Do you remember the *iodine clock*? Your teacher *controlled* a chemical reaction.
You can control reactions too!

Your teacher will give you a time deadline for this task. You will need to work quickly. Everyone must be involved.
Will your group get the best result by the test deadline?

Your task is to:

Make a 100 cm³ sample of carbon dioxide gas in 60 seconds.

Your timing should be exact. You need to make *exactly* 100 cm³ of gas. (Not 99 cm³ or 101 cm³.)
Try to get as close to this volume as possible in the 60 seconds.
Use limestone and acid to make your carbon dioxide.

acid – eye protection

Your teacher will ask you to demonstrate your reaction to the other groups at the test deadline.

1 You have been asked to make a jelly for your brother's birthday party. The party is due to start in 2 hours! What will you do to dissolve the jelly cubes in water as quickly as possible?

2 Jan wants to make and collect a sample of carbon dioxide gas.
a) Draw a labelled diagram of the apparatus she could use.
b) Describe 2 uses of carbon dioxide.

3 When magnesium ribbon reacts with an acid, it makes hydrogen gas. Ben measured the volume of gas given off every minute.

Time (minutes)	Volume (cm³)
0	0
1	20
2	35
3	45
4	50
5	52
6	52
7	52
8	52

a) Draw a graph of these results.

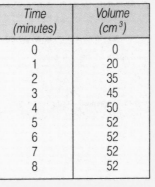

b) After how many minutes had the reaction finished?
c) What volume of hydrogen had been collected at the end of the reaction?
d) When was the reaction fastest . . . near the start or near the end?
e) Give one way of making this reaction happen faster. Try to explain this by writing about *particles*. What happens to the particles of magnesium and acid in the reaction?

Things to do

125

Catalysts can help

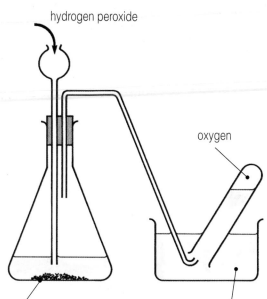

hydrogen peroxide

oxygen

manganese(IV) oxide

water

▶ Think about how to make a reaction happen faster.
Make a list of the different ways to do this.

There is another way to change the rate of a reaction.
Catalysts can help.

You've already used **catalysts** in science before.
Do you remember making *oxygen*?

Manganese(IV) oxide is a catalyst. It makes the hydrogen peroxide decompose (break down) quickly.

A catalyst is a substance which changes the rate of a chemical reaction. A catalyst is not used up during the reaction.

Using catalysts

Have you ever looked *closely* at a packet of washing powder**?**
The powder in the photograph is like lots of others you can buy.
The list of ingredients on the box is interesting.
The powder contains **enzymes**.
On the box it says:

Ingredients	Function
Enzymes	Break down stains containing proteins e.g. blood, milk and stains containing fats e.g. body soils, cooking fat, etc.

So what are enzymes**?** You might remember them from topic 16 (page 47).
Enzymes are biological catalysts.
Enzymes in the washing powder help to break down stains quickly.
We say that the fat and protein in stains are **digested** by enzymes.

Enzymes help to make these.

Yeast contains enzymes.

▶ Look back in this topic.
Write a few lines to describe the reaction which used yeast.
(Hint: without the yeast, you'd still be waiting for the bubbles!)

Which is the better catalyst?

Hydrogen peroxide can decompose (break down). It makes water and oxygen.

hydrogen peroxide ⟶ water + oxygen

You already know that manganese(IV) oxide is a catalyst for this reaction. It makes hydrogen peroxide decompose quickly. But Leon has another idea:

> I want to make oxygen quickly. I think liver could be a better catalyst than manganese(IV) oxide.

Plan an investigation to test Leon's idea.

Show your plan to your teacher. Then do the investigation.

⚠️ Hydrogen peroxide can cause burns. Wear eye protection.

Things to do

1 Copy and complete:
A catalyst is a substance which . . . . the rate of a reaction.
A catalyst is not . . . . during the reaction.
Liver is a catalyst to make . . . . gas.
An . . . . is a biological catalyst.

2 Think about the reaction to make oxygen. Liver is a catalyst.
Does the temperature of the liver affect the rate of the reaction?
Plan an investigation to test this.

3 Catalysts can let reactions happen at a lower temperature than usual. Why do people in industry think catalysts are a good idea?

4 Find out about catalytic converters. (Your local car showroom may let you have some leaflets.)
a) How do catalytic converters work?
b) What are the advantages of using them?
c) Are there any disadvantages?

Even though the use of unleaded fuel has cut exhaust emissions, new environment laws have been enforced. Every petrol-powered engine comes with a highly efficient 3-way catalytic converter as standard. This greatly reduces harmful emissions.

Questions

1 Lee and Asha were looking at ways to prevent rust. Their teacher asked them which is the best method.

Who is right?
Plan an investigation to find out.

2 The UK chemical industry is the 5th largest in the world.
It makes many different products.
Draw a bar-chart or a pie-chart of these data:

Products made	Percentage of the industry (%)
Fertilisers	8
Organic materials	12
Inorganic materials	7
Soaps and toiletries	9
Pharmaceuticals	24
Plastics and rubbers	5
Paints and varnishes	8
Dyes and pigments	4
Specialised chemical products	23

3 A new iron and steel works is to be built in the UK.
Imagine you are in charge. You have to decide **where** to build it. What things would you need to think about? Write a list.

4

Coal is a better fuel than wood.

a) Do you think this statement is correct?
b) What makes a good fuel?
   What tests could you do to see which fuel is better?

5 a) Make a list of the main gases in the air.
b) Design an experiment to find out how much oxygen is in the air. (The diagram opposite might give you a clue about one possible way.)

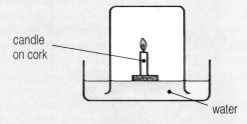

candle on cork

water

6 Draw a poster to summarise everything you have learnt about reactions in this topic.
Use colour and pictures to make your poster interesting.

7 Read the memo and imagine you are Andrew.
Write at least half a page to advise Liz.

8 If you use more catalyst, does the reaction go faster?
Plan an investigation to find out.

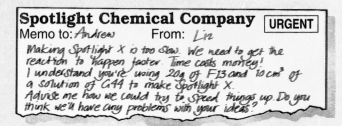

Spotlight Chemical Company URGENT
Memo to: Andrew   From: Liz
Making Spotlight X is too slow. We need to get the reaction to happen faster. Time costs money!
I understand you're using 20g of F13 and 10 cm³ of a solution of G44 to make Spotlight X.
Advise me how we could try to speed things up. Do you think we'll have any problems with your ideas?

Electricity

23

What has your brain got in common with lightning, fire-alarms, and computers?
They all use electricity.

In the world today, there are electronic systems everywhere – at home, at school, at work, in hospitals. They help to make our lives easier.

In this topic you will be inventing and building electronic systems.

Electric charges

Have you ever rubbed a balloon until it can stick to a wall**?**

Have you ever rubbed a pen or comb until it picks up bits of paper**?**

We say the objects are **charged**.
They have **static electricity**.

Investigating charges

You can use strips of plastic, made of:
- acetate (a clear plastic),
- polythene (a grey plastic).

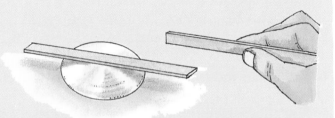

Rub one of the strips on a dry cloth, and then balance it on a watch-glass, so that it can spin easily:

Rub another strip on the cloth and then bring it near one end of the balanced strip. What happens**?**

Copy the table:
Then find out what happens in each case.
Fill in your table with the words **attract** or **repel**:

What pattern do you find**?**

		Strip in my hand	
		acetate	polythene
Strip balanced on watch-glass	acetate		
	polythene		

We say that the acetate gets a positive charge (+).
The polythene gets a negative charge (–).

We say an object is **un**charged when it has got **equal** amounts of positive charge and negative charge.
The charges are **balanced**. It is **neutral**.
Count the number of charges on these neutral objects:

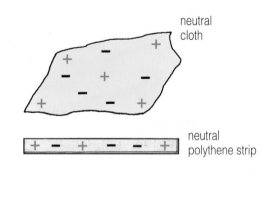

neutral cloth

neutral polythene strip

When you rub the strip with the cloth, the charges become **unbalanced**. Some negative charges have moved, from the cloth to the polythene strip:

Count the number of charges on each object now:
What do you find**?**

Draw a similar diagram to show what happens when you rub an acetate strip.

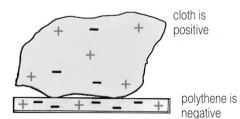

cloth is positive

polythene is negative

Moving charges

Look at a Van de Graaff generator:
It is a machine for making static electricity.
The moving belt carries electric charges up to the dome.

Your teacher will show you some experiments with it.
For each one, describe carefully what you see, and explain
why it happens.

a Some small pieces of paper are put on the dome, and then
the machine is turned on.
What happens, and why?

b Some threads are fastened to the dome, or a piece of fur is
put on the dome.
What happens, and why?

c A brave volunteer with loose dry hair stands on a plastic
sheet and touches the dome.
What happens, and why?

d Why does the volunteer have to stand on a plastic sheet?
Is plastic a conductor?

e A large metal ball is brought near the dome. What
happens, and why? Where do you see this in nature?

f While the dome is sparking, a metal pin is pointed at the
dome.
What happens? Why do you think this happens?

g How is this idea used to protect tall buildings from
lightning?

h In the dark, a neon lamp or a fluorescent light is held near
the dome.
What happens? What do you think this means?

---

1 Copy and complete:
a) If charges are the same sign (+ or –)
they . . . . each other.
If charges are different signs then they
. . . . each other.
b) In a neutral object, the . . . . are
balanced. In an object which is
negatively-charged there are more . . . .
charges than . . . . charges.
c) Electricity can pass through a . . . . but
not through an . . . .

2 Why might it be dangerous to use an
umbrella in a storm? Where should you **not**
shelter in a storm?

3 Try to explain each of these:
a) In dry weather, people walking on nylon
carpets may get a shock when they
touch a radiator or a metal door handle.
b) If you clean a dusty mirror or window
with a dry cloth, on a dry day, it can be
even more dusty the next day.
c) When you pull 'cling film' off the roll
it clings to your hands.
d) Old 'cling film' does not cling so well.
e) Cassette-cases, records, and other
plastic objects soon become dusty.

4 Which do you sense first: thunder or
lightning? Why is this?

Things to do

Current affairs

▶ Look at this circuit diagram:

a What does each symbol stand for**?**
b What happens if the circuit is not complete**?**

We say that the ammeter and the bulb are **in series**.
Tiny **electrons** are moving through the wires.

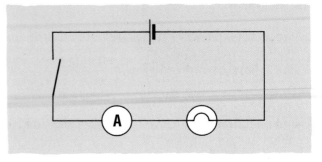

c Draw a circuit diagram for a battery (cell) and 2 bulbs in series.
d What can you say about the brightness of the 2 bulbs**?**
What can you say about the current through the 2 bulbs**?**
What happens if one of the bulbs breaks**?**

e What is an **insulator?**
f Draw a diagram of a circuit you could use to test if an object is an insulator or a conductor.

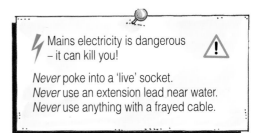

Mains electricity is dangerous – it can kill you!

Never poke into a 'live' socket.
Never use an extension lead near water.
Never use anything with a frayed cable.

g Now draw a circuit diagram for a battery and 2 bulbs **in parallel**.
h What happens if one of the bulbs breaks**?**

A circuit contains a battery, a switch, and 3 bulbs labelled X, Y, and Z. Bulbs X and Y are in series, and bulb Z is in parallel with X. The switch controls only bulb Z.

i Draw a circuit diagram of this.
j When all the bulbs are lit, which one is the brightest**?**

Investigating resistance

Connect up this circuit:

Take care to connect the ammeter the correct way round (+ of ammeter nearest to + of the battery).

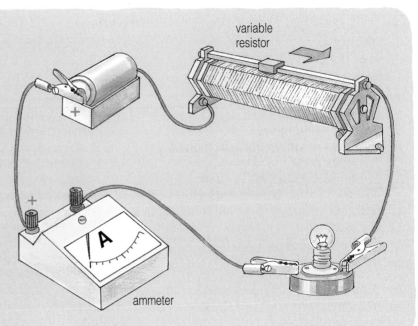

variable resistor

ammeter

k Are the components in series or in parallel**?**

l The current goes through only part of the variable resistor. On the diagram, is this the yellow part or the blue part**?**

m What happens when the slider is moved to the right on the diagram**?** Try it. **Why** does this happen**?**

n If the bulb is not lit up, does it mean there must be no current in the circuit**?** Try it.

o Draw a circuit diagram of your circuit.

Making a fire-alarm

A **bi-metallic strip** is made of 2 metals fastened together.
When it is heated, the metals get longer. They *expand*.

But one metal expands more than the other.
What happens to the strip? Try it. Explain what you see.

clamp

brass

iron

Design a fire-alarm using a bi-metallic strip.

- Think about how you can use it to ring a bell or light a warning lamp.
- Draw a circuit diagram of your design.

Ask your teacher to check your circuit. Then build it and test it.

Sketch a drawing of your fire-alarm. Describe how it works.
How could your design be improved?

How could you change your design to warn you if something got too cold?
Can you think of a use for this?

You have used the bi-metallic strip as a **sensor**.

How big is the push?

Plan an investigation to see how the **voltage** across a bulb depends on the **number of batteries** in the circuit.

- Draw a circuit diagram.
- How can you record your results?
- Predict what you think will happen.
- Ask your teacher to check your plan, and if you have time, try it.
  Was your prediction correct?

voltmeter

---

1 Copy and complete:
a) If the same current goes through two bulbs, then the bulbs are in . . . .
b) If the current splits up to go through two different paths, then we say the paths are in . . . .
c) A good conductor has a . . . . resistance. An insulator has a . . . . resistance.
d) An ammeter measures the . . . . in a circuit, in . . . . or A.
e) A battery pushes . . . . round a circuit. The size of the push is measured in . . . . , by using a . . . .

2 Draw circuit diagrams of these:
a) A battery (cell) and a switch connected to 2 bulbs in series, with a voltmeter across one of the bulbs.
b) Two batteries (cells) in series, connected to 2 bulbs wired in parallel, an ammeter to measure the total current taken by the bulbs, and a switch to control each bulb.

3 Draw a circuit diagram to show how 3 lamps can be switched on and off separately but dimmed all together.

Things to do

Electronic systems

In the last lesson you made a fire-alarm system.
It was made from several components.
You joined them together to make a **system**, to do a job.

All systems are made up of 3 basic parts:

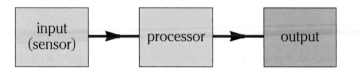

For your fire-alarm,
The **input** was the heat which made the bi-metallic strip
bend. The bi-metallic strip is a **sensor**.

The **processor** was the circuit which was completed by
the contact.

The **output** was a bell ringing (or a bulb lighting up).

Here is another system, a radio:
The input is the radio signal coming into the aerial.
The processor is the amplifier inside the case.
The output is the sound coming out of the loudspeakers.

▶ Write down   i) the input and   ii) the output, for each
   of these systems:
a   a door-bell
b   a coffee-machine
c   your calculator
d   putting your hand in hot water!

Electronic systems are now very common.
▶ Read this short story:

   Lisa saw a fan-club advertised on **teletext** but she
   lost part of the address. So she went to the library
   and used a **computer** to search for the address on a
   database. Then she wrote a letter to the fan-club
   using a **word-processor**. She checked it on the
   VDU and stored it on **disc**. Then she sent her letter
   by **electronic mail**, using a **modem**.
   The fan-club sent her a membership form by **fax**.

e   Explain each of the bold words, in as much detail as
   you can.

In your group, discuss the effects of this new
technology.
• How has it changed life over the last 20 years?
• What are its advantages and disadvantages?
• How do you think electronics will change life
  over the next 20 years?

Analogue or digital?

The hands on a clock are always on the move. They move on a continuous scale round the dial. This is an **analogue** system.

In a **digital** clock, the time is shown in steps. It is only changed every minute.

f Look at these photos, and decide which things are analogue and which are digital.

g What are the advantages and the disadvantages of these 2 methods of display?

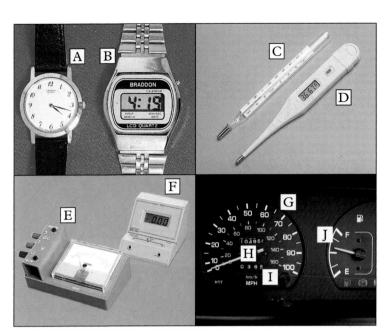

Electronic systems are often digital systems. They often have only 2 states: OFF and ON. For example, your fire-alarm had only 2 states: it was OFF or ON. A light switch is either OFF or ON.

Binary logic

A system with only 2 states is also called a binary system. When it is OFF it has a binary code of **0** (zero). When it is ON it has a binary code of **1**. These are also called **logic 0** and **logic 1**.

▶ Copy this table and complete it:

Binary code	Switch	Lamp	Bell	Current flowing	Answer to a question	Logic
1		ON			YES	1
0	OFF		silent	NO		0

Sending signals

The Morse Code is made up of dots and dashes:

h Is it an analogue or a digital signal?

i Draw a circuit diagram of a system to send and receive Morse Code from someone in another room.

j Write a message in Morse Code and send it to a friend. Can they de-code it correctly?

A	• —	J	• — — —	S	• • •
B	— • • •	K	— • —	T	—
C	— • — •	L	• — • •	U	• • —
D	— • •	M	— —	V	• • • —
E	•	N	— •	W	• — —
F	• • — •	O	— — —	X	— • • —
G	— — •	P	• — — •	Y	— • — —
H	• • • •	Q	— — • —	Z	— — • •
I	• •	R	• — •		

1 Copy and complete:
a) All systems have . . . . basic parts: input, ... , and . . . .
b) An . . . . signal can have a continuous range of values. A signal which can be only ON or OFF is called a . . . . signal.

2 Are you in favour of video-phones? Write a short story about a home of the future that uses a lot of new technology.

3 Write down i) the input and ii) the output, for each of these systems:
a) a burglar alarm b) an electric guitar
c) a torch d) automatic doors in a shop
e) a TV set f) you hearing your name.

4 Are these analogue or digital?
a) the volume control on a radio
b) traffic lights
c) the school bell.

Things to do

Your eye is a **sensor**. It detects light and gives an input to your system.

a   What other sensors do you have in your body?

In your fire-alarm, the sensor was the bi-metallic strip. It detected a change in temperature.

Here are some more sensors:

Movement

A **micro-switch** starts and stops a current.
It senses the movement and produces an electrical signal.

A **tilt switch** contains 2 wires and a blob of mercury, as shown. Mercury is a liquid, and a conductor.

b   What can happen if this switch is tilted?

c   Design a burglar alarm that will go off if a desk lid or a car bonnet is lifted up. Draw a circuit diagram.

d   Where else could you use a tilt switch sensor?

Magnetism

A **reed switch** is a magnetic switch. It detects when a magnet comes near.
See the experiment on the opposite page.

Moisture

A **moisture sensor** has 2 wires close together but not touching. If a rain-drop lands so that it touches both wires, a current can flow. This is because rain-water is a conductor.

e   Who might want a moisture detector?

Sound

A **microphone** detects sound and makes an electrical signal.

Light

A **light-dependent resistor** is called an **LDR**.
See the investigation on the opposite page.

f   Where would a light sensor be useful?

Temperature

A **thermistor** is a useful temperature sensor.
See the investigation on the opposite page.

g   Where would a temperature sensor be useful?

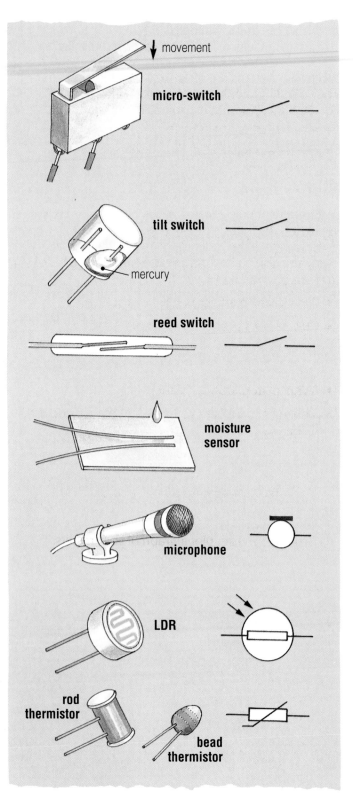

movement

micro-switch

tilt switch

mercury

reed switch

moisture sensor

microphone

LDR

rod thermistor

bead thermistor

Looking at a reed switch

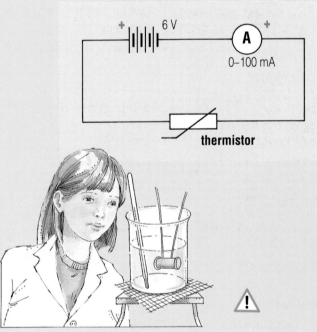

reed switch

- What happens as you bring the magnet near? At what distance does this happen?
- Use a magnifying glass to look closely at it.
- Devise a burglar alarm for a window, using a reed switch.

Investigating an LDR (light-dependent resistor)

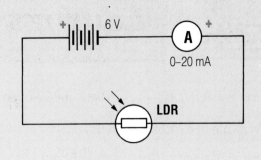

6 V

A
0–20 mA

LDR

- What happens as you vary the amount of light on the LDR?
- When does most current flow? When does the LDR have least resistance to the current?
- How does the current depend on the colour of the light? Which colour is best?

Investigating a thermistor

Use this circuit to investigate how the **current** depends on the **temperature** of the thermistor.

- How many readings will you take?
- How will you record your results?

6 V

A
0–100 mA

thermistor

- What pattern do you find?
- When does the thermistor pass the most current? When does it have the least resistance?

- Design a thermometer for a car. It should show the driver the temperature of the engine.

- If you have time, plot a graph of your results.

---

1 Copy and complete:
a) A reed switch is a . . . . switch.
   It is switched on when a . . . . comes near to it.
b) An LDR is a . . . . -dependent . . . .
   The brighter the . . . . that shines on it, the more . . . . can flow through it.
c) A thermistor is a . . . . sensor.
   The hotter it is, the more . . . . can flow through it.

2 Make a list of the electronic sensors that you have in your home.

3 Suggest a sensor for each system:
a) To tell you if it is raining outside.
b) To tell you if the baby upstairs cries.
c) To tell a pilot that he is taking off too steeply.
d) To tell a blind person that it is time to close the curtains at night.

Things to do

a Every system has 3 parts. What are they?

b If you pick up a very hot object, what is your output signal?

c What was the output from the fire-alarm that you made?

This page tells you about some of the **output** devices that you can use in electronic systems.

Light

You have already used filament lamps in your circuits.

LED lights are often used in electronics.
LED stands for **l**ight-**e**mitting **d**iode.
LEDs are usually red or green. They use much less current than filament bulbs.

d Where have you seen LEDs used?

Movement

An electric **motor** uses electricity to make movement.

e What is the name for the movement energy in a motor?

f Make a list of the motors in your home.

Sound

A sound can be made by a bell, a buzzer, or a loudspeaker.

g When do you use a loudspeaker?

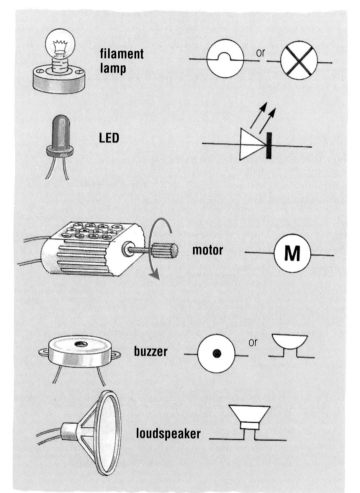

Using a relay

In some circuits, we might want to use a small current to switch on a larger current.

We can do this by using a **relay**. It contains an electromagnet.
Look carefully at this diagram:

There are 2 circuits here.

h When a small current flows in the blue circuit, what happens to the iron core?

i How does this change the red circuit?

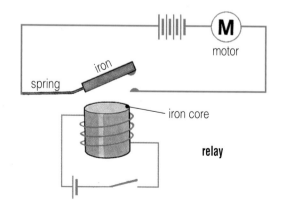

Making connections

Use an electronics kit to investigate inputs and outputs.
If the kit is like the one shown here, try to ignore the
middle part of the board for now.

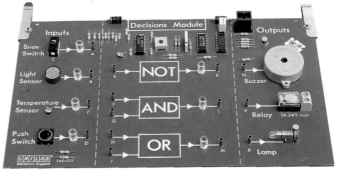

▶ Look at each of the inputs.

An input is ON when its red LED lights. This is also
called logic **1**.
Find out how to make each input go ON and OFF.

▶ Connect a wire from the **slide switch** to the **lamp**:

How do you make the lamp go ON and OFF?

Now try each of the sensors with the lamp.
And then with the buzzer.

switch   lamp

Look closely at the **relay** while you switch it ON and OFF.
What do you see?
How can you use the relay to switch a motor ON and OFF?

▶ You can make a **moisture sensor** like this:

What happens when you connect the 2 wires with a
damp cloth?
You could use this as a wet-nappy detector!

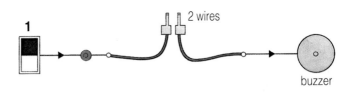

2 wires   buzzer

Now design circuits for **j** to **p**.
Test each design, and then draw a labelled diagram of it.
The first one has been done for you.

light sensor   buzzer wakes
farmer at dawn

j An alarm to wake up a farmer at dawn:

k A fire-alarm for a blind person.

l A burglar alarm using a pressure pad.

m A burglar alarm for a drawer.
  (Hint: it is usually dark inside a drawer.)

n A hot-breath detector:

o To switch on a fan when the room gets too hot. If possible,
  connect a motor and a battery to your relay.

p To tell a blind person when to stop pouring from a tea-pot.

1 Copy and complete:
a) LED stands for:
b) In a relay, a small . . . . can switch on a
  large . . . .

2 What do these symbols mean?
a)   b)   c)

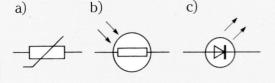

3 Draw a connection diagram (like the
ones above) for each of these:
a) An alarm if a chip-pan gets too hot.
b) An alarm that detects a burglar's torch.
c) To tell you if it is raining outside.
d) To open a greenhouse window if it gets
  too hot.
e) To switch on a water-pump if a ditch
  fills with water.
f) To call a nurse if a patient's temperature
  is too high.

Things to do

Making decisions

A system has 3 parts. Can you remember what they are**?**

The middle part of a system is often a **logic gate**.
It is an electronic 'chip' that can make decisions.

Logic gates act rather like doors or garden gates. They
only let you through if you open them in the correct way.

NOT gate

Connect a push switch directly to a buzzer:
What happens**?**

Now include a NOT gate, like this:
What difference does it make**?**

Do the same with the other input sensors.

How can you make the buzzer come on when it is
NOT warm**?** (It is cold.)

Then when it is **NOT** light**?** (It is dark.)

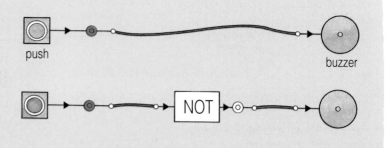

> The output signal is **NOT** the same as the input signal.

AND gate

Connect a slide switch and a push switch to the
AND gate, like this:

What do you have to do to get the buzzer to sound**?**

What do you notice about the red LEDs**?**
Remember: ON is a logical **1**, OFF is a logical **0**.

How can you make the buzzer come on when it is
light **AND** warm**?**

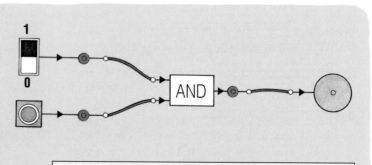

> The output signal is **1** (ON) only if:
> the first input is **1**   **AND**   the second input is **1**.

OR gate

An OR gate opens in a different way from the other
gates.
Connect the OR gate like this:

What are the rules to make the buzzer work now**?**

How can you make the buzzer come on when it is
warm **OR** the push switch is pressed**?**

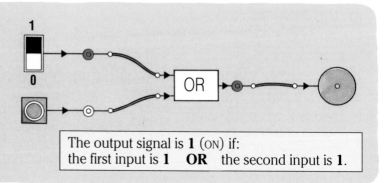

> The output signal is **1** (ON) if:
> the first input is **1**   **OR**   the second input is **1**.

Solving problems

Problem: Baby Jane is in an incubator. It is very important that the nurse is warned if the baby gets cold. Can you help?

Solution: Warm up your thermistor with your finger and make this system:

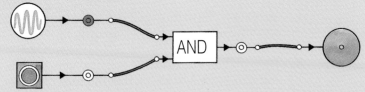

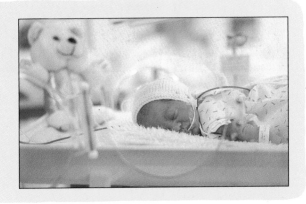

What happens when the thermistor cools down?
In the hospital, where would you put each part of the system?

Problem: Mr Smith likes to go to bed early, and doesn't want his doorbell ringing at night. Can you help?

Solution: Try this system:

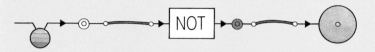

What happens at night? Why?
The system is making the decision for you.

Problem: Jenny is designing a new push-chair.
Her idea is that a buzzer will warn you if you let go of the handle (a push switch) without putting on the foot-brake first (a slide switch). Can you help?

Solution: You'll need 2 gates for this:

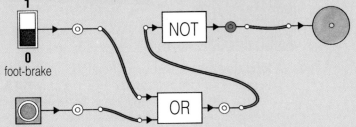

What happens if the foot-brake is off (logical 0) and you let go of the handle switch? Why?

---

1 Copy and complete:
a) NOT, AND and OR are three . . . . gates.
b) In a NOT gate, the output is . . . . the same as the input signal.
c) In an AND gate, the output is ON (logical 1) only if the first input is 1 . . . . the second input is 1.
d) In an OR gate, the output is 1 (ON) if the first input is 1 . . . . the second input is 1.

2 Draw system diagrams (like the ones above) to solve these problems:
a) To switch on a street-light when the sky goes dark.
b) To open the window of a greenhouse if it is warm **and** it is day-time. (You can use a relay to switch on a motor.)
c) To sound an alarm if the temperature in a freezer gets too high **or** if the door is left open (so that light is shining in).

Things to do

These photographs show some situations where logic gates can help.

▶ For each one,

- decide which gate or gates you will need,
- build the system and test it,
- draw a system diagram (like on page 141),
- say where you would put the sensors.

⚠ Do not attempt any of these ideas with mains electricity!

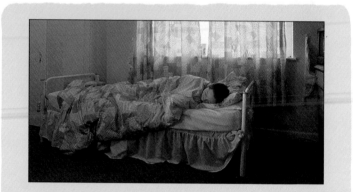

1 Design an alarm that will wake you up if it is day-time **and** your room is warm.

2 Then add an OR gate and a test switch, so that you can press it to check that the buzzer works.

3 Design a fire-alarm which has an OR gate with a test switch, so that you can press it to check that the buzzer works.

4 Can you add a slide switch and an AND gate, so that it 'enables' or 'disables' the fire-alarm system.

5 A kind dog owner has put a heater in the kennel. The heater (a lamp) should come on only if it is cold *and* the dog is in the kennel (pressing down a pressure switch). When the kennel warms up, the heater must switch off. Can you help?

6 Mrs Brown lives alone. She is nervous about answering the door at night. She'd like a porch light that comes on automatically when it is dark. Can you help?

7 Mrs Brown's bills are adding up. She'd like you to add a slide switch, so that she can switch off the light when she goes to bed. Can you help?

8 This machine is fitted with a safety guard to protect the user. When the safety guard is correctly in position it breaks a beam of light.

The machine should start (by switching on a relay) only when the safety guard is in position and the start button is pressed. Can you help?

9 Can you add a wire, so that the buzzer warns when the safety guard is not in place?

10 A burglar alarm. This treasure must be protected by 2 sensors. It is standing on a pressure switch (a push switch), and there is a light sensor under it also. If any light gets to the light sensor, or if the push switch is released, the alarm must sound. Can you help?

11 The pump of a central-heating system should be switched on if the system is 'enabled' (by a slide switch) and if it is cold. When the temperature rises, it should be switched off (by a relay).

A lamp should show you when the pump is on.

12 Mike is a keen gardener. He wants an alarm to warn him when it is freezing outside, so that he can protect his prize cabbage. Can you help?

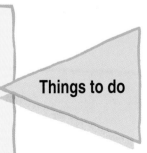

13 Now he wants the alarm to work only at night.

14 And now he wants a slide switch to disable the system, so that he can get back to sleep after covering up his cabbage. Can you help him?

---

1 Draw system diagrams to solve these problems:
a) To switch on a fan in a hospital if a patient's temperature is too high or if the nurse decides it should be on.
b) To allow a car to be started only if the ignition switch is on and the driver's seat-belt is fastened.
c) To warn you if a tropical fish tank is dark or cold.

2 Paula wants a porch light that will come on only when it is dark and somebody is standing on the doormat.
a) Draw a system diagram of the logic gates that Paula will need to use.
b) Sketch a porch showing where you would put the different components.
c) How can you add a slide switch so that Paula can put on the light at any time?

Things to do

Questions

1   Carl was wearing a woollen jumper over a nylon shirt.
   a) When he started to take off the jumper, it crackled.
      Why is this?
   b) When he had taken off the jumper, he found it was
      attracted to his shirt. Why?

2   The diagram shows a circuit with two 2-way switches.
   Wire C can be connected to either A or B.
   This circuit is often used for the lighting on a staircase.
   a) In the diagram, is the lamp on or off?
   b) Describe carefully how the circuit works, using the letters
      on the diagram in your answer.
   c) What are the advantages of this circuit?

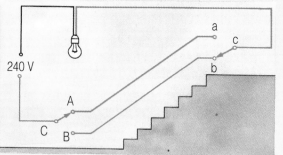

3   Keung and Donna are talking about batteries.
   They each have a hypothesis:
   Keung says, "The HP2 cell is larger and so it will
   make the bulb brighter."
   Donna says, "The HP7 cell will make the bulb just
   as bright but the cell won't last as long."
   a) Who do you think is right?
   b) Plan an investigation to find out who is right.

4   Look at the display on your calculator.
   a) Why is it called a seven-segment display?
   b) Why is it called a digital display?
   c) If only segments a, b, c are ON, what number is displayed?
   d) Using logic 1 = segment ON, this could be written in code as
      1110000. What number would be displayed if the code is
      1101101?
   e) Write down the code to display a number 5.

5   a) The warden of a nature reserve wants an alarm that will
      warn her if anyone goes near the nest of a rare bird. Design
      a system to do this, making it as fool-proof as possible.
   b) Then she asks for a way to monitor the temperature of the
      bird's eggs from her office. Draw a circuit diagram for her.

6   a) A laundry needs an alarm to warn if the hot-water tank
      starts to cool down during the day. Design a system for
      this and include a test switch.
   b) Design a system to warn you when the soil of your house-
      plant becomes dry. You wouldn't want it to warn you
      during the night.

Plants at work

24

Can you imagine what life would be like without plants?

We use plants for food, fuel, building materials and medicines. Plants also take waste carbon dioxide out of the air and make oxygen for us to breathe.

The great energy trap

Plants can't eat like animals do.
So how do they get their food?

▶ Write down some ideas about how you think plants feed.

Plants make their food from simple substances.
But to do this they need energy.
Where do you think this energy comes from?
What part of the plant do you think traps this energy?

Plants make food by the process **photosynthesis** (photo = light
and synthesis = to make). To make food they need:

- carbon dioxide from the air
- water from the soil
- light energy trapped by **chlorophyll**.

CARBON DIOXIDE + WATER $\xrightarrow[\text{CHLOROPHYLL}]{\text{SUNLIGHT}}$ SUGAR + OXYGEN

▶ Write down the answers to these questions:

a What food do plants make themselves?

b Which gas is made during photosynthesis?

c How do animals use this gas?

Oxygen bubbles

Jill and Emma had seen fish tanks with air bubblers. They knew
that these were to put oxygen into the water for the fish to breathe.
Jill read that if there is lots of pondweed and the tank is well lit, air
bubblers aren't needed.

Jill and Emma set up an investigation to see the effect of light on the
pondweed.
They put some pondweed in a test-tube of pond water.
They placed a lamp at different distances from the test-tube and
counted the number of bubbles of gas produced in a minute.

Their results are shown in the table:

Distance of lamp from pondweed (cm)	10	20	40	Lamp off
Number of bubbles produced per minute	15	7	4	2

d Which gas do you think was produced by the pondweed?

e How could you prove this?

f Do you think there is a pattern to these results?

g Emma thought that the lamp might also warm up the
pondweed and not make it a fair test. How could you improve
the design of their investigation to avoid this?

Testing a leaf for starch

Most of the sugar made in the leaves of a plant is changed to starch.
You can test for this starch with iodine.
If the leaf turns blue-black with iodine then starch has been made.

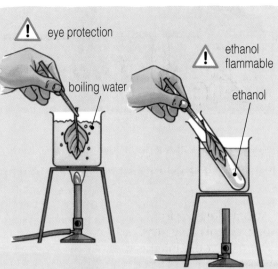

eye protection

ethanol flammable

boiling water

ethanol

turn Bunsen off

- Dip a leaf into boiling water for about 1 minute to soften it.
- Turn off the Bunsen burner.
- Put the leaf into a test-tube of ethanol. Stand the test-tube in the hot water for about 10 minutes.
- Wash the leaf in cold water.
- Spread the leaf out flat on a petri dish and cover it with iodine. What colour does the leaf go?

iodine

h Why was it important that you turn off the Bunsen burner when you were heating the ethanol?

i What was the leaf like after you heated it in the ethanol?

j Was there any starch in the leaf that you tested?

Into the light

If starch is present we can say that the leaf has been making food.
Plan an investigation to see if a plant can make food without any light.

Stripes and patches

Not all leaves are green all over.
Some have white and green patches, others stripes.
If you have time, try testing one of these leaves for starch.

k Which parts do you think will go blue-black?

Remember to draw a picture of what your leaf looks like at the start to show which parts are green.

Things to do

1 Copy and complete:
Plants make their food by . . . . They use . . . .
. . . . from the air and . . . . from the soil. They also need a green substance called . . . .
which traps . . . . energy. The food that is made is sugar and it is changed to . . . . in the leaf. The waste gas made is called . . . .

2 Plants are important because they provide:
a) food   b) fuel
c) building materials   d) medicines.
Find examples of plants that provide each of these things.

3 Explain why a) and b) are vital to our survival.
a) Plants use up carbon dioxide.
b) Plants release oxygen.

4 Joseph Priestley found that a lighted candle in a jar soon went out. He put a plant in the jar and shone light on it for a week. He found the lighted candle now burned much longer. Can you explain his experiment?

Leaves

24b

What do you think is the most common colour in nature?

Plants are green because they contain **chlorophyll**.

▶ Look round the room and out of the window.
  Write down the names of 5 plants that you see.
  In which parts of the plants do you think there is most chlorophyll?

Leaf design

▶ Look closely at both sides of a leaf.

a Which side is the darker green?

b Which side do you think has most chlorophyll?

c Why do you think this is?

▶ Write down some words to describe the shape of your leaf.
  A leaf's job is to absorb as much light as it can.
  How do you think its shape helps it to do this?

Looking inside a leaf

Look at a section of a leaf under your microscope.
Focus onto the leaf at low power.
Now carefully change the magnification to high power.
Can you see any of the parts labelled in the diagram on your slide?

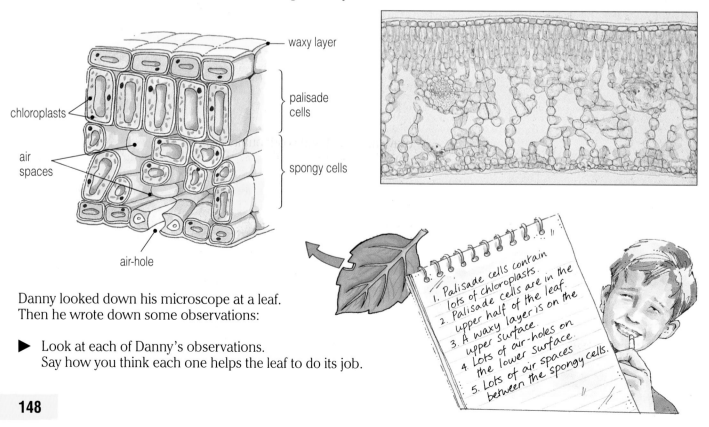

Danny looked down his microscope at a leaf.
Then he wrote down some observations:

▶ Look at each of Danny's observations.
  Say how you think each one helps the leaf to do its job.

148

Holey leaves

Try dropping a leaf into a beaker of boiling water.
Which surface do bubbles appear from?
Why do you think this is?

⚠️ eye protection

Gases from the air pass into and out of a leaf through **air-holes**.

Paint a small square (1 cm × 1 cm) on the underside of a leaf with nail varnish.
The nail varnish will make an imprint of the leaf surface.

Wait for it to dry completely. (While you are waiting, you can set up your microscope.)
Carefully peel off the nail varnish with some tweezers.
Put it on a slide with a drop of water and a coverslip.
Observe and draw 2 or 3 air-holes at high power.
Repeat for the upper surface of your leaf.

Write down your conclusions.

Form 8GMW wrote down some ideas for investigations:

Do leaves in different positions on a plant have the same number of air holes?

Do different species of plant have different numbers of air holes?

Do the same species of plant grown in different places have the same number of air holes?

If you have time, plan **one** of these investigations, check your plan with your teacher, then try it out.

Things to do

1 Match each of the leaf parts on the left with the job that they do on the right:

air-holes   carry water up from stem
palisade cells   allow gases to pass into and out of leaf
spongy cells   contain many chloroplasts
waxy layer   contain many air spaces
veins   prevents too much water being lost

2 Do plants of the same species always have the same leaf area?
Plan an investigation to compare the leaf area of the same plants found in light and shady conditions. Try to explain any differences that you find.

3 Leaves have a large surface area to absorb light. Lay a leaf onto graph paper, draw around it and work out its area by counting the squares. Can you work out the total leaf area of the plant?

4 Leaves are thin so gases can get in and out easily. But they can also lose water and then they droop.
Look closely at a leaf and say what helps to stop them from drooping.

Plant growth

A lot of our food comes from plants.
Think about what you have eaten over the last 24 hours.

▶ Write down the food that you think came from plants.

Plants for food

Farmers try to grow enough food for us to eat.
They try to give their crops the best conditions for growth.

a   Make a list of things you think plants need to grow well.

▶ Look at the photographs of lettuces growing in a greenhouse:
Those in A are growing in air which has more carbon dioxide
than those in B.

b   Which do you think would sell for the best price?

c   Can you explain why these lettuces are bigger?

d   If you were growing a crop in a greenhouse, how could you:
    i)   increase the amount of time that the plants are in the light?
    ii) keep the plants at the right temperature?

Fertilisers

Plants also need chemicals called **nutrients** for healthy growth.
These are usually found in the soil.
They are taken up in small amounts by the roots of the plant.
If the soil does not contain enough nutrients, then the farmer adds
more as **fertilisers**.

▶ Look at the table showing the effects of fertilisers on wheat:

e   Which fertiliser gives the biggest increase in growth?

f   Which nutrients are found in fertiliser B?

g   Which nutrient do you think is the most important for wheat
    growth?

Fertiliser	Nitrogen added	Phosphorus added	Potassium added	Wheat yield (tonnes per hectare)
none	×	×	×	1.70
A	✓	×	×	3.80
B	×	✓	✓	2.00
C	✓	✓	✓	7.00

NPK fertilisers contain:

- Nitrogen (N), for general growth,
- Phosphorus (P), for healthy roots,
- Potassium (K), for healthy leaves.

The proportions of nitrogen, phosphorus and potassium (N : P : K)
are shown on the fertiliser bag.

h   Which nutrient is missing from the fertiliser in the picture?

i   Another fertiliser is called 25 : 5 : 5. What do you think this means?

NPK fertilisers

Give more, grow more

We can use liquid fertiliser (plant food) to grow good house plants.

Plan an investigation to see how the growth of some duckweed depends on the fertiliser.

You could look at the effects of either
or
- different fertilisers
- different strengths of the same fertiliser.

- What are you going to measure to show growth?
- What things will you need to keep the same if it is to be a fair test?
- What are you going to change?
- How long will your investigation last?
- How often will you take measurements?

Show your plan to your teacher before you try it out.

Where there's muck, there's growth

Why do you think gardeners put horse manure on their roses?

Animal waste is broken down by microbes.
Some bacteria get nitrogen out of the waste.
Plants take up the nitrogen through their roots and use it for growth.

j How else do you think manure **improves** the soil?

k Why do you think some gardeners prefer to use chemical fertilisers?

▶ Find out what you can about **organic gardening**.

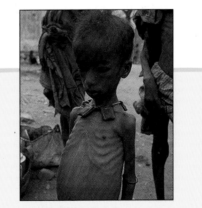

1 Copy and complete:
To grow, plants need . . . . from the air, water from the . . . . , and sunlight. They also need nutrients from the soil. These include N (. . . .), P (. . . .) and K (. . . .). If the . . . . does not contain enough nutrients, the farmer adds . . . .

2 Animal manure and compost are both good for plant growth.
Explain why you think this is true.

3 What nutrients are found in NPK fertilisers?
If a fertiliser has an NPK value of 10 : 5 : 10, what does it mean?

4 We are often told that enough grain is grown to feed everyone in the world. So why do you think it is that people are starving?
Use the following clues to explain why there is hunger in the world:
a) transport b) wars c) food mountains
d) pests   e) drought.

Things to do

Plant plumbing

The roots of a plant grow into the soil.

a   Write down what you think the roots are for.

b   What do you think would happen to a plant if it had no roots?

▶ Look carefully at some different roots.

c   Write down ways in which they look different from the rest of the plant.

seed
young shoot
root hairs
young root

What do you think roots would look like under the microscope?
You would see lots of tiny hairs called **root hairs**.
The root hairs take up water from the soil.

d   How do you think their shape helps them to do this?

Up to the leaves

Your teacher will give you a piece of celery.
This has been standing in water containing a dye.

Carefully cut off about 1 cm as shown in the photograph.
Make an accurate drawing of the inside of the celery.
Colour the parts where you can see the dye.

The dyed water is carried in tiny tubes called **xylem** (sigh-lem).

Carefully cut out a 2 cm length of your xylem.
Look at it with a hand-lens.
Describe what you see.

Stem support

Water passes into a plant through the roots.
It then passes in the xylem up the stem to the leaves.

e   What else do you think the stem does?

f   What would happen to the leaves and flowers if there was no stem?

There are other tiny tubes in the stem called **phloem** (flow-em).
Look at the diagram:

g   Write down what you think the phloem does.

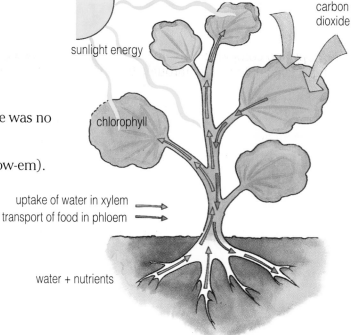

carbon dioxide
sunlight energy
chlorophyll
uptake of water in xylem
transport of food in phloem
water + nutrients

Waterways

How do you think water travels up to the tops of tall trees?
As water is lost from the leaves, more water moves up from the stem.

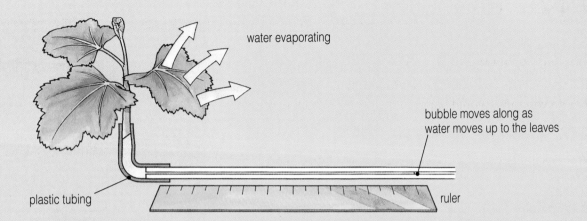

water evaporating

bubble moves along as
water moves up to the leaves

plastic tubing

ruler

You can measure how quickly water is moving up the stem to the
leaves with this apparatus. Your teacher will show you how to set it up.

Measure the distance travelled by the bubble every minute.

Plot your results on a line-graph with axes like this:

Now repeat your experiment, but this time either:

- cover the shoot with a clear polythene bag
- place a fan near your apparatus.

or

Plot your results on the same line-graph as before.

Now try to explain why you have got different results.

Distance
moved by
bubble
(cm)

Time (min)

If you have time:
Look at a microscope slide of the inside of a stem.
Try to find the xylem and the phloem.

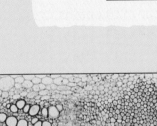

phloem

xylem

Things to do

1 Copy and complete:
Water is taken into a plant through its . . . .
It is then carried up the . . . . in tiny tubes
called . . . . Water is lost from the plant in the
. . . . and more . . . . is drawn up the . . . .
from the roots.
Food made in the . . . . passes to the rest of
the plant in tubes called . . . .

2 Leaves lose water, just like washing on
a line.
In what sort of conditions do you think
leaves would lose i) most water?
ii) least water?
Try to explain this.

3 Weeds have strong roots that anchor
them in the soil.
Design a piece of apparatus to test how
strong the roots of some common weeds
are.

4 How strong is a stem? Suppose you are
given 2 different pieces of stem each 10 cm
long. Plan an investigation to find out which
is the stronger.
Include a diagram of the apparatus you
would use.

Pollination

Many people give flowers to friends and relatives.
But how many realise that flowers are the plant's reproductive system?
Plants have male and female sex cells just like animals.
The male sex cell in a plant is called a **pollen grain**.
The female sex cell in a plant is called an **ovule**.

▶ Write down some words that describe what flowers are like.

A closer look

You can use a hand-lens to look at your flower in more detail.
Cut the flower in half like the one in the picture.
See if you can find all the parts labelled in the picture.

Many flowers have male and female parts.
The female parts are called the **carpels**.
Look at the picture:

a What is each carpel made up of?

The male parts are called the **stamens**.

b What is each stamen made up of?

Use a hand-lens to look closely at these parts in your flower.

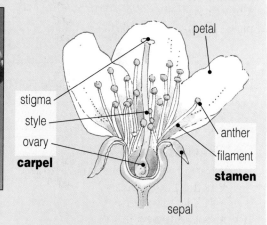

Make a flower poster

Take a new flower and carefully remove all the parts with tweezers.
Start on the outside with the sepals. Then work inwards removing the petals, stamens and carpels.
Arrange the parts in a line, one under the other, in your book.
Stick them down neatly with sellotape and label the parts.

Looking at pollen

Inside each **ovary** the ovules are made.
Inside each **anther** lots of pollen grains are made.

Take a stamen from one of your flowers.
Scrape some pollen off the anthers with a mounted needle.
Place the pollen onto a microscope slide.
Add a drop of water and place a coverslip over it.
Look at your pollen grains at high power under your microscope.
Draw 2 or 3 of your pollen grains.
If other groups have pollen grains from different flowers, look at these.

c Write down your ideas of what these pollen grains are like.

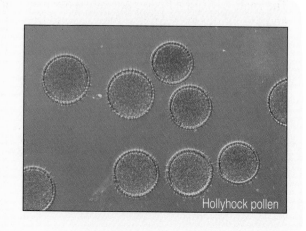

Hollyhock pollen

Carried by insects

Pollination is the transfer of pollen from the anthers of a flower to the stigma.

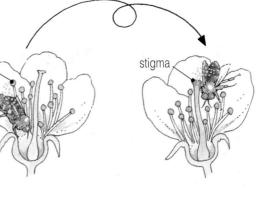
anther   stigma

d   At what time of year are most flowers out?

e   At what time of year are insects such as bees and butterflies out?

Insects are a great help in carrying pollen from one flower to another.
But first, the flowers have to attract the insects to them.

f   Write down 3 ways in which flowers can attract insects.

▶ Look at the picture showing **insect pollination**.

g   Why do you think the bee reaches down into the first flower?

h   How does the bee carry pollen to the second flower?

i   Where does the bee leave pollen in the second flower?

Blown by the wind

Not all flowers need insects to pollinate them.
Many flowers like grasses and cereals rely on the wind to carry their pollen to another flower.

▶ Look at a wind-pollinated flower with a hand-lens.

j   How does it compare with your insect-pollinated flower?
  • Is it brightly coloured?
  • Does it have a scent?
  • Does it have **nectar**?

k   Why do you think this is?

l   What do you think its pollen is like?
  • Is the pollen sticky?
  • Is it light or heavy?
  • Is much pollen produced?

m   Why do you think this is?

▶ Make a table to show all the differences that you have found out between insect-pollinated flowers and wind-pollinated flowers.

---

1   Copy and complete:
The male parts of a flower are called the . . . . and are made up of 2 parts: the . . . . and the . . . . The female parts of a flower are called the . . . . and are made up of the . . . . , the . . . . and the . . . . Pollination is the transfer of . . . . from the anthers of one flower to the . . . . of another.

2   Cross-pollination is when pollen is transferred to separate flowers. What do you think is meant by **self-pollination**? How could it take place?

3   Did you know that 1 in 10 people suffer from **hay fever**?
Find out what the symptoms are.
When do you think people suffer most?
Find out what the **pollen count** is and what sort of weather conditons can affect it.

4   Copy and complete these half-sentences:
a)   Some flowers attract insects because they have . . . .
b)   Wind-pollinated flowers do not need colour because . . . .
c)   Insect-pollinated flowers have sticky pollen because . . . .
d)   Wind-pollinated flowers make lots of pollen because . . . .

Things to do

Making seeds

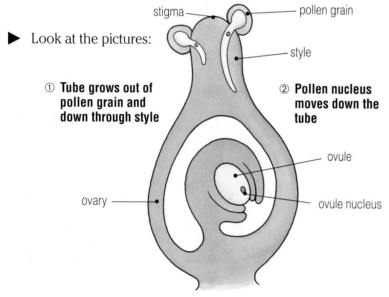

Where do you think seeds come from?
Where could you find seeds?

▶ Write down some of your ideas.

After the flowers have done their job they die.
Left behind will be the **fruits** containing the seeds.

First things first

A **pollen nucleus** must join with the **ovule nucleus**.

a What do we call this joining together of a male nucleus and a female nucleus?

Once it is fertilised, the ovule grows into a **seed**.

▶ Look at the pictures:

stigma — pollen grain

style

① **Tube grows out of pollen grain and down through style**

② **Pollen nucleus moves down the tube**

ovule

ovary

ovule nucleus

③ **Pollen nucleus joins with ovule nucleus. Fertilisation takes place and a seed will form.**

b Where do you think the pollen grain comes from?
c What happens to a pollen grain after it lands on a stigma?
d How does the pollen nucleus reach the ovule nucleus?

After fertilisation, the ovary changes to form the **fruit**.

e What do you think the fruits are like in each of these plants:
 i) grape? ii) oak tree? iii) pea? iv) tomato?

The seeds grow from the ovules inside the fruit.

f Can you remember the 3 parts of a seed from Book 7? Write them down.

g What 3 things do you think seeds need before they will grow (**germinate**)?

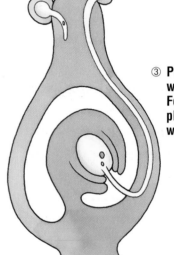

Scattering seeds

Seeds are often scattered over a wide area.

h Why do you think this is important**?**

Seeds can be scattered by wind, animals and by being flicked out of pods.

▶ Look at these pictures:

sycamore

apple

dandelion

sweet pea

burdock

strawberry

gooseberry

poppy

acorn

For each one, write down how you think the seeds are scattered.

Seed fall

"Seeds that fall slowly have a better chance of being carried further by the wind" said the man in the gardening programme.
Do you think that this is true**?**

Plan an investigation to find out how slowly different seeds fall.

* Think carefully about what apparatus you will need.
* What measurements are you going to make**?**
* Remember to make it a fair test.

Show your plan to your teacher, then try it out.

---

1 Write out the following sentences in the correct order to describe how plants reproduce:
A Pollen nucleus joins with ovule nucleus.
B Pollen grain lands on stigma.
C The fertilised ovule becomes a seed.
D The pollen grain grows into a pollen tube.
E The pollen nucleus travels down the pollen tube to the ovule.

2 What do we mean by fertilisation?
In a flowering plant, what takes the place of:
a) the sperm?   b) the egg?
c) the fertilised egg?

3 What happens to each of the following after fertilisation:
a) the flower?   b) the ovule?
c) the ovary?

4 a) Why don't seeds germinate in a gardening shop?
b) What do they need to grow?
c) Copy the drawing of a germinating seed and label it using these words:

> new leaves   seed coat   new root
> new shoot   food store

Things to do

Questions

1 Label the parts of the leaf using the following words:
palisade cells, air spaces, epidermis, air-holes, waxy layer, spongy cells, chloroplasts.
Write down: A = chloroplasts, etc.

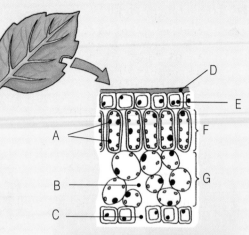

2 Sanjit shone different amounts of light on some pondweed. He recorded the number of bubbles of gas given off by the pondweed per minute:

Units of light	Number of bubbles per minute
1	6
2	14
3	21
4	24
5	26
6	27
7	27

a) Draw a line-graph to display his results.
b) How many bubbles of gas would you expect the plant to make at
   i) 2.5 units of light?   ii) 8 units of light?

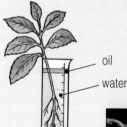

3 Farmer Jenkins put lots of fertiliser on his fields in the autumn. The following summer the river nearby was full of water weeds. Some of the weeds died and started to rot. This took oxygen out of the river.
a) Why do you think the water weeds grew so much?
b) What do you think happened to the fish in the river?
c) How would you solve this problem?

4 The apparatus shown in the diagram was used to measure how much water was lost from the leaves in 24 hours. The apparatus was weighed at the start and at the end of the experiment.
a) Explain how you think the apparatus works.
b) What do you think the results would show?
c) What do you think the oil is for?

oil
water

5 Potatoes are full of starch.
But the starch is made in the leaves.
So how do you think the starch gets into the potatoes in the soil?

6 Stems grow up and roots grow down.
But does it matter which way up a seed is planted in the soil?
Plan an investigation to find out.

7 a) Why do you think that woodland plants like primroses and wood anemones flower in the spring?
b) Why do you think hazel 'catkins' make lots of pollen?
c) Why do you think it is important that seeds get far away from the parent plant?

Index